The Recovery of

Gold

from Secondary Sources

The Recovery of Gold from Secondary Sources

Editor

Syed Sabir

King Saud University, Saudi Arabia

Imperial College Press

ICP

Published by

Imperial College Press
57 Shelton Street
Covent Garden
London WC2H 9HE

Distributed by

World Scientific Publishing Co. Pte. Ltd.
5 Toh Tuck Link, Singapore 596224
USA office: 27 Warren Street, Suite 401-402, Hackensack, NJ 07601
UK office: 57 Shelton Street, Covent Garden, London WC2H 9HE

Library of Congress Cataloging-in-Publication Data
Names: Sabir, Syed, 1970–
Title: The recovery of gold from secondary sources / [edited by] Syed Sabir
     (King Saud University, Saudi Arabia).
Description: New Jersey : Imperial College Press, 2016.
Identifiers: LCCN 2016013084 | ISBN 9781783269891 (hc : alk. paper)
Subjects: LCSH: Electronics--Materials--Recycling. | Gold--Recycling.
Classification: LCC TD799.85 .R424 2016 | DDC 622/.3422--dc23
LC record available at http://lccn.loc.gov/2016013084

British Library Cataloguing-in-Publication Data
A catalogue record for this book is available from the British Library.

Desk Editors: Herbert Moses/Mary Simpson

Typeset by Stallion Press
Email: enquiries@stallionpress.com

Printed in Singapore

Contents

Acknowledgments   vii

About the Editor   ix

1   Introduction   1
    S. Syed

2   Leaching of Gold from the Spent/End-of-Life
    Mobile Phone-PCBs using "Greener Reagents"   7
    Jae-chun Lee and Rajiv R. Srivastava

3   Electroless Displacement Deposition of Gold from Aqueous
    Source — Recovery from Waste Electrical and Electronic
    Equipment (WEEE) using Waste Silicon Powder   57
    Kenji Fukuda and Shinji Yae

4   Adsorption of Gold on Granular Activated Carbons
    and New Sources of Renewable and Eco-Friendly
    Activated Carbons   95
    Gerrard Eddy Jai Poinern, Shashi Sharma, and Derek Fawcett

5   Development of Novel Biosorbents for Gold
    and Their Application for the Recovery of Gold
    from Spent Mobile Phones   143
    Katsutoshi Inoue, Manju Gurung, Hidetaka Kawakita,
    Keisuke Ohto, Durga Parajuli, Bimala Pangeni,
    and Shafiq Alam

**6   Environmentally Friendly Processes for the Recovery
     of Gold from Waste Electrical and Electronic Equipment
     (WEEE): A Review**   173
     Isabella Lancellotti, Roberto Giovanardi, Elena Bursi,
     and Luisa Barbieri

**7   Study on the Influence of Various Factors
     in the Hydrometallurgical Processing of Waste
     Electronic Materials for Gold Recovery**   197
     I. Birloaga and F. Vegliò

Index   221

Acknowledgments

The Editor gratefully acknowledges the College of Engineering Research Center and the Deanship of Scientific Research, King Saud University, for supporting this book.

About the Editor

Dr. Syed Sabir is a Research Professor at the King Saud University, Riyadh, Saudi Arabia and visiting Professor at Riyadh College of Technology, Saudi Arabia. His research interests include hydrometallurgy (Au and Ag), hydrophobic nanomaterials, hydrophobic biomass, carbon dioxide capturing, water pollution, and atmospheric degradation of materials, etc. He has published over 50 peer-reviewed research papers in international journals, including three invited reviews, as well as four invited books, one of his papers *"Recovery of Gold from Secondary Sources — A review"*, ranked 1st in the Top 25 Hottest Articles, for Hydrometallurgy — 2012–2016 full year (January–December). He has over 500 citations and received Hydrometallurgy Top Cited Author for 2012 to 2016, Awarded (Hydrometallurgy, Elsevier Ltd. New York, USA). He is a reviewer of more than 40 international journals and a research proposal reviewer of various international research institutes. He is a member of copious international scientific societies and has several international collaborations. His books and research papers are globally used as standard reference sources.

Chapter 1
Introduction

S. Syed

Since the dawn of the human history, gold has incalculable applications, especially as the prime sources of currency entry in the economics of any state [1]. Gold is a lustrous metal of great beauty that has enthused great works of art which have persisted to remain unchanged for thousands of years. This immutable, most noble of noble metals has principally been prized for its stability, an extraordinary ductile and malleability, and it is still a celestial demand in jewelry, high-tech industries and medical applications because of their distinctive physical and chemical properties [2]. Conversely, we have lately understood that gold, when divided to form particles at the nanoscale, can take on new properties; these fascinating properties have driven a new "gold rush". The recovery of gold continues to grow in importance as gold deposits all over the world are shrinking and depleted, and it can be predicted that the natural sources of gold will be exhausted by the end of the year 2030 and the market for consumer electronics continues to grow (ACS reports 30 mg of gold in iPhone 6, requiring over 2,300 kg of Au to cover their Apple's recent sales) [3]. This prediction results in an intensive search for new gold sources.

S. Syed
Chemical Engineering Department, King Saud University
P.O. Box 800, Riyadh-11421, Saudi Arabia
e-mail: sabirsyed2k@yahoo.com, ssabir@ksu.edu.sa

Studies in the field of gold containing waste have grown in considerable measure following the fact that the sum and rate of spent gold material generation is increasing every year. It is therefore mandatory to study alternatives to recycle and reuse gold waste, to extend the life cycle of gold, and to reduce the rate of extraction of natural resources to supply the industrial demand for gold. Thus, the gold component has become a contender for recovery. Obviously, the process of recovery has logic only if the cost of recovery is considerably less than the value of the gold. Moreover, limitations levied on waste disposal and draconian environmental regulations demand economical, viable, and eco-friendly technologies [4]. Here, we focused on gold, which can generally be recovered by hydrometallurgical and bio-hydrometallurgical processes and can be extracted from leachants via adsorption, electro-less displacement deposition, bio-sorption, chemical precipitation, cementation, electro-wining, ion exchange, solvent extraction, coal–oil agglomeration (CGA), coagulation and reduction, etc. [5,6].

1.1 Secondary Gold Bearing Raw Materials

In spite of the fact that a considerable quantity of the world's production of gold comes from the direct processing of gold ores, the recovery of gold from spent sources remains a significant and potentially increasing supply source, and henceforth the operations intricately warrant a technology review [7]. All industrial and domestic gold containing secondary resources are divided into liquid, metallic, and non-metallic sources. The natural sources of gold are various wastes, e.g. spent gold plating solutions, gold plating drag-outs, seawater, copper anode slimes, residues of gold extraction factories, scrap jewelry, spent multi-component alloys–solders, spent metallo-ceramic substrate, plated rejected components, spent dental and orthopedic materials, mine tailings, various slags, sludge from gold electrolysis, waste from grinding and polishing sections, spent catalysts, porcelain scraps, glass bangles scraps, loaded carbon and filter, adsorbed ion exchange resins, mobile phone scraps, and electronic waste. The latter is also called as waste of electrical and electronic equipment (WEEE) or electronic waste (e-scrap), etc. [8–11]. A scrap containing gold may ascend

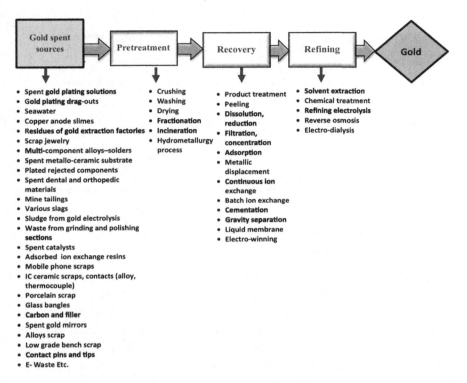

Fig. 1 Basic processing flow sheet

from a diversity of sources and is available in various forms; processing of such scraps is, therefore, intricate. Nevertheless, the basic processing technology is shown in Fig. 1 [2]. The pretreatment process comprises washing, crushing, separation, and incineration depending on the nature of scraps, and the metal achieved is of low purity. In the recovery process, such low purity metals are transformed into crude gold by repeated dissolution, filtration, concentration, and reduction. Crude gold are subjected to the refining process to improve their purity to over 99.9%. The flowchart is shown in Fig. 2 [2]. The recovery conditions can only be established after a complete characterization of the gold waste and scrap, i.e. only after thoroughly understanding the gold containing waste, should the recovery processes and routes be chosen and applied.

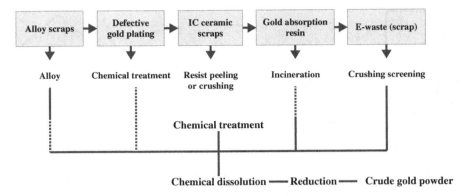

Fig. 2   Flowchart of gold recovery from gold spent sources

1.2 Scope of the Book

The aim of this book is to summarize the comprehensive overview of the gold containing secondary sources. In addition, processing techniques for the recovery of gold from spent sources were described. Therefore, Chapter 2 has been devoted to the discussion on the mobile phone-printed circuit board (MP-PCB) containing a significant amount of gold (up to 440 g/t Au), approximately 100 times higher than that of the usual rich-alluvial ores (2–4 g/t Au). To recover such a captivating amount of the gold from the PCBs of end-of-life mobile phones, its leaching in an aqueous solution is of primary concern during the hydrometallurgical processing. The stringent environmental regulation worldwide has attracted great attention towards the usage of the "Greener Reagents" for the sustainable metallurgy of gold. Usability of the "Greener Reagents", such as thiosulfate, thiourea, iodine, and *in situ* generated chlorine with the process limitations and challenges is being discussed. The techno-economic advantages of gold leaching from the MP-PCBs with these reagents than their applicability in the leaching of ore bodies are also examined. Although the chemistry of gold leaching from MP-PCBs and alluvial ores with the lixiviants employed is almost similar, both the raw materials exhibit different mechanisms because of their divergent nature, which is also being explained.

Chapter 3 is focused on the gold recovery using an electro-less displacement reaction, i.e. cementation. This is a simple but efficient method to refine gold ore into pure metal, and has been practically used in industries for over 100 years. The cementation is also useful for noble metal recovery from WEEE, urban mines in other words. The principle of electro-less displacement deposition and conventional cementation processes of gold are briefly reviewed, and then the novel method to recover gold from such secondary source as a plating solution and the WEEE using waste silicon powder are described.

Chapter 4 discusses the several novel adsorption gels which were prepared from various biomass such as lignin, polysaccharides, and persimmon tannin all of which were found to exhibit a very high selectivity and loading capacity for gold (III). Here, gold was successfully recovered as aggregates of fine particles of metallic gold. This noteworthy phenomenon was attributed to the reductive adsorption mechanism caused by many hydroxyl groups on the surface of these gels. Through effective manipulation of such unique characteristics, these novel adsorbents can be expected to be employed in recovering gold contained in various solutions at dilute concentrations directly as metallic gold. This was exemplified in the recovery of gold from PCBs derived from spent mobile phones using the adsorption gel prepared from persimmon tannin. After the pretreatments involving dismantling and crashing followed by incineration, the ash of incinerated PCBs was leached using hydrochloric acid into which chlorine gas was blown. It was confirmed that gold can be successfully recovered from such leach liquor using a column packed with this gel.

Chapter 5 outlines the major contemporary gold refining processes, currently using activated carbons and new sources, such as macadamia nutshell activated carbons, manufactured from waste products produced by this agricultural sector.

Chapter 6 illustrates the e-waste issues, then being focused mostly on the actual hydrometallurgical processing technologies of waste printed wiring boards that are known as main components of all electronic devices. Therefore, a short literature overview of all current technologies was undertaken and, according to it, the main factors that influence the recovery process of Au with different solvents during the leaching procedures will be identified and discussed in this chapter.

Chapter 7 presents the state of art of environmentally friendly processes for the recovery of gold from WEEE with a particular attention to sulphur compounds and chloride-based leachants. A case study on the recovery of gold from waste computer PCBs using a chloride-based leachant capable to etch the metals that support the gold (copper, nickel, and iron) was also presented in detail.

Although extensive work has been reported on the processing techniques for the recovery of gold from various spent sources, this book would serve as a new vista for scientists, researchers, engineers, and postgraduate students.

References

[1] Konyratbekova SS, Aliya B, Ata A (2015) Non-cyanide leaching processes in gold hydrometallurgy and iodine-iodide applications: A review. *Mineral Processing and Extractive Metallurgy Review*, **36**: 198–212.

[2] Syed S (2012) Recovery of gold from secondary sources — A review. *Hydrometallurgy*, **115–116**: 30–51.

[3] Moyer M (2010) How much is left? *Scientific American*, **303**(3): 74–81.

[4] Syed S (2006) A green technology for recovery of gold from non-metallic secondary sources. *Hydrometallurgy*, **82**: 48–53.

[5] Cui J, Zhang L (2008) Metallurgical recovery of metals from electronic waste: A review. *Journal of Hazardous Materials*, **158**(2–3): 228–256.

[6] Gramatyka P, Nowosielski R, Sakiewicz P (2007) Recycling of waste electrical and electronic equipment. *Journal of Achievements in Materials and Manufacturing Engineering*, **20**(1–2): 535–538.

[7] Ailiang C, Zhiwei P, Jiann-Yang H, Yutia MA, Xuheng Liu, Xingyu C (2015) Recovery of silver and gold from copper anode slimes. *The Journal of The Minerals, Metals & Materials Society*, **67**(2): 493–502.

[8] Potgieter JH, Potgieter SS, Mbaya RKK, Teodorovic (2004) A small-scale recovery of noble metals from jewellery wastes. *Journal of The South African Institute of Mining and Metallurgy*, **104**(10): 563–571.

[9] Li J, Xu X, Liu W (2012) Thiourea leaching gold and silver from the printed circuit boards of waste mobile phones. *Waste Management*, **32**(6): 1209–1212.

[10] Chmieleskil AG, Urbanski TS, Migdal W (1997) Separation technologies for metals recovery from industrial wastes. *Hydrometallurgy*, **45**(3): 333–344.

[11] Kolobov GA, Bredikhin, VN, Chernobaev, VM (1993) Collection and Processing of Recycled Ferrous Metals. Moscow: Metallurgy'93, p. 289.

Chapter 2
Leaching of Gold from the Spent/ End-of-Life Mobile Phone-PCBs using "Greener Reagents"

Jae-chun Lee and Rajiv R. Srivastava

2.1 Introduction

Gold has the traditional, financial, and technological values worldwide for its widespread area of applications. With the development of human civilization, gold became a medium of financial trade, and nowadays, it is an integral part of the modern cutting-edge technologies in electronic equipment and medical diagnostics. Such characteristics present gold as a precious metal whose market demand is consistently high in contrast to depleting gold concentration in ore bodies from 2 g/t to 0.4 g/t [1]. To fulfill the demand of 3923.7 t gold, only 3114.4 t gold could be supplied in the year 2014 while mine production was on a record high [2]. A big gap between the demand and supply of gold results in price hikes, unstable markets and therefore makes it crucial to explore alternative routes. Electronic gadgets, which contribute to ~10% (equivalent to 389 t) of

J.-c. Lee
Mineral Resources Research Division,
Korea Institute of Geoscience and Mineral Resources (KIGAM)
e-mail: jclee@kigam.re.kr

R. R. Srivastava
Resources Recycling, Korea University of Science and Technology (UST)

Table 1 Composition of critical metals and their value
shares in various e-waste [3–6]

Item particulars	Weight share, ppm			Value share, %		
	Au	Ag	Pd	Au	Ag	Pd
TV board	20	280	10	27	8	8
PC board	250	1000	110	68	5	15
Mobile phone	440	1380	210	82	5	21
Portable audio	10	150	4	14	4	3
DVD player	15	115	4	37	5	5
Calculator	20	260	5	74	7	4

total gold consumption [2], after completion of their life span, can be
recycled in material flow systems and has great potential to lead towards
sustainability.

However, the contents of gold in each of the electronic gadgets differ
considerably (Table 1) from 3 g/t to 440 g/t [3–6], and thus, every recycling
process of electronic waste (e-waste) cannot be economically viable due to
the higher processing costs than the yield. Among the electronic gadgets,
mobile phones contain the highest amount of gold (up to 440 g/t that
contributes to >80% value share of the whole gadget as shown in Table 1)
that widely depends on the origin, model, and even manufacturer of the
phone [3–6]. This analysis reveals that as an average, 1 g of gold can be
recovered by the recycling of 30–40 mobile phones that is equivalent to
the processing of per ton of alluvial deposits [7]. Besides this, each ton of
mobile phones has 130 kg copper, 3.5 kg silver, and 0.14 kg palladium to
make the process more profitable to recycle the waste mobile phones than
other e-waste.

The recycling of discarded mobile phones looks more attractive by
considering only its printed circuit boards, i.e. mobile phone-printed circuit
boards (MP-PCBs), as shown in Fig. 1, which is the main reservoir of
gold (on an average 980 g/t Au) in the form of thin coated layer due to the
excellent characteristics of electrical conductivity [8], low contact electrical
resistance, and outstanding corrosion resistant. A typical journey of this
communication device to golden gadgets is pictorially presented in Fig. 2
[5–8]. The morphology of different sections observed by SEM clearly shows

Fig. 1 A typical picture of MP-PCBs showing the upper layer coated with gold

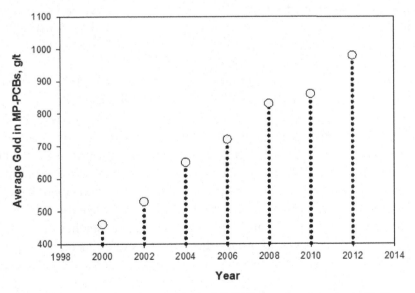

Fig. 2 Increasing gold content in the MP-PCBs as a function of year [5–8]

(a)   (b)   (c)

Fig. 3 A typical representation of the PCBs of scrapped mobile phone: (a) as such, (b) after size-cutting, and (c) SEM image showing the layerwise metallic composition [6]

four layers of materials in an MP-PCB from its outer to inner side (Fig. 3(c)), including the Au–Ni alloy, nickel, copper, and plastics [6]. The energy-dispersive spectrum analysis shows that the outer layer of Au–Ni alloys consists of >90% of gold. Therefore, other metals that are not exposed much come into contact directly with the lixiviant at initial leaching stages that can keep the gold leaching unaffected. The recycling of MP-PCBs seems to be more attractive when compared with the mining, crushing, and grinding costs of the primary deposits.

There are several recycling routes of gold recovery from the e-waste such as pyro-, pyro+hydro-, and hydrometallurgical processes. A typical scheme for the unit operations mainly involved in the recycling routes is shown in Fig. 4 [9]. It can be understood that a pyrometallurgical route comprises costlier high temperature incineration and smelting processes, which require sophisticated flue-gases cleaning system to capture and clean the hazardous and toxic gases such as furans, dioxins, and carbon monoxides. In general practice, therefore, it is unviable to treat the PCBs to only recover the gold, and often, a fixed amount is being fed into copper smelter as a mixed charge with the copper concentrates. The metallic values, including copper and gold, is recovered by following the established industrial process of copper extraction, thereby gold along with other precious metals are settled down in anodic chamber as slimes during electro-refining process. The quantity of collected anode slimes from a typical copper electro-refinery varies in the range of 2.5–25 kg/t of cathode processed, containing up to 1,200 g/t Au [10] which undergoes several hydrometallurgical processing

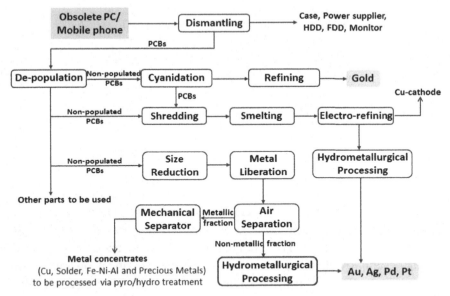

Fig. 4 A schematic flowsheet for the processing of e-waste recycling (including the pyro and hydrometallurgical treatments of waste MP-PCBs [9]

stages to recover the gold and other metals from them. A significant amount of precious metals that remained in slag during smelting can be considered to be disadvantageous for the entire operation of gold recovery, whereas in only hydrometallurgical processing of e-waste, the metals are being solubilized into the acidic/ammoniacal/cyanide solutions followed by several separation and purification steps. Notably, to employ the hydrometallurgical processes, the costlier size reduction followed by the efficient physical beneficiation processes is a prerequisite to proper liberation and enrichment of metals into aqueous solution from the PCBs [9, 11].

The hydrometallurgical processing is wider in commercial practice of gold extraction in comparison to the pyro-processes because of more accurate, predictable, and controlled attributes of the aqueous processing. In particular for gold recovery, a leaching step with aqua regia solution or prior cyanide leaching of e-waste (including the MP-PCBs) is employed widely. In the hydrometallurgy of gold, leaching with royal acidic solution of aqua

Table 2 The adverse effect of cyanide on human
health as a function of HCN concentration [13]

HCN exposure (in ppb)	Effect on human health
Up to 500	None
500–1,000	Flushed, tachycardia
1,000–2,500	Obtunded
2,500–3,000	Coma
above 3,000	Death

regia and alkali cyanide solution are in application from the 8th century and the last decade of the 19th century, respectively, due to the ease in formation of soluble species of gold with the two reagents and obtainment of higher recovery [12]. The cyanidation is easier to operate, and even, it can be employed in the forefront of pyrometallurgical operation for gold recovery (Fig. 4). But the environmental hazards produced in both the processes as toxic NO_X and HCN gases need to be strict on safety measures, and contamination with natural water resources is of much concern in recent years. The adverse effect of HCN on human health (shown in Table 2) may cause death [13]. There is a necessity, therefore, to be replaced by the economic and environmental friendly process in gold recovery. A lot of unorthodox work is being done. Bacterial leaching can be an alternative to it [14], but the slow kinetics and adaption of bacteria are some practical challenges. In this view, the focus is still on the chemical leaching with "greener lixiviants". With a prime objective of reducing the environmental hazards along with resource recovery during the recycling of waste MP-PCBs, the application of "green lixiviants" is vital.

This chapter is devoted to the investigations and developments in gold leaching from the waste MP-PCBs using the reagents which are considered to be non-toxic and greener alternative of the cyanidation and aqua regia leaching. Factors affecting the gold recovery, viz. characteristics of the particular reagent, process chemistry, extraction mechanism, and separation from different metallic constituents, etc., are being discussed here with the limitations and challenges of each process.

2.2 Technological Background for the Need and Scope of Greener Alternatives

Most of the gold production is still achieved by cyanidation till date. The toxicity of cyanide is a major concern (Table 2); however, cyanidation is a significantly robust process [10]. Plenty of works have been reported and patented with the well-defined chemistry and extraction mechanism to form the stable gold-cyanide complex ions which hardly give operational problem [10, 15]. The carbon-in-pulp (CIP) process has reduced the operational cost by eliminating the solid–liquid separation needed in Merrill–Crowe process. But, cyanidation is unable to effectively leach out the gold from carbonaceous refractory ores and shows lack of selectivity in terms of copper. In these cases, the consumption of cyanide remains high and tailing discharge of highly concentrated cyanide solution in comparison to the permitted level as low as <0.1 ppm [16] becomes more difficult. Similar to this, when we consider the specific gold leaching from the materials like PCBs either via direct hydrometallurgical or pyro+hydrometallurgical routes, the presence of carbonaceous matter in solution coming from the detached hydrocarbon of the PCB resins causes "preg-robbing" of gold in CIP. This adversely affects the gold recovery and is fair enough to make the process unviable.

The powerful oxidizing environment created by the aqua regia can vigorously dissolve the maximum of metals (including the base metals), therein PCBs present the tedious work to evaporate the free nitrates, neutralization of excess acid followed by several separation and purification operations in downstream processing to recover the gold, ultimately increasing the overall cost [17]. Direct treatment of aqua regia leach liquor without any neutralization work in downstreaming has been claimed [18]. Though the developed process is simpler in operation, it does not encounter the hazards generated in the form of NO_X during upstream leaching process.

While searching for a greener option to replace the cyanide/aqua regia leaching, not only the environmental cause for the human health and ecological hazards need to be taken care of, but also the technological part should relate to the reagent employed in leaching. After all, the metal extraction/recycling make sense only if the process is technologically sound enough to lower the cost than the value of the metal produced.

Otherwise, it will be difficult to practically replace the cyanidation. The cost, consumption, recyclability, and commercial availability of the reagent itself, exhibition of fast kinetics, ease in application to give high efficacy, and understanding of the extraction chemistry are some of the vital factors which should be taken into consideration to present an alternative instead of considering only the environmental stewardship. In this case, the literal sustainability of gold can be achieved with the application of "greener lixiviant".

2.3 Greener Reagents and their Applicability in Gold Recovery from the Spent MP-PCBs

Numerous research activities on e-waste recycling, including the spent mobile phone, are going on worldwide, and more than two dozens of reagents have been identified potentially to be "greener" [19]. However, only a few have been significantly attracted due to the technological compulsions that have been derived in the above section. Among them, the sulfur-based compounds and halides are mainly adopted to investigate in gold recovery process from the spent MP-PCBs. Interestingly, chlorine–chloride leaching was in commercial practice in the 19th century to recover gold from alluvial ores, but its application diminished with the cyanidation in 1889 [10]. The highly corrosive nature and their transportation was mainly responsible for that which is now being handled by the proper designed engineering, use of less corrosive materials, and *in situ* electro-generated chlorine (the process developed in KIGAM, South Korea for gold recovery from the spent MP-PCBs) [6, 10, 20, 21].

The evaluation of various indexes of several leaching methods has been done by using the analytic hierarchy process [22]. The result of the economic, environmental, and technological reliability analysis of various processes is shown in Table 3 [22, 23]. Based on this, while focusing on gold leaching from spent MP-PCBs using "greener lixiviant", the entire discussion can broadly be divided into two parts:

➢ Leaching with sulfur-based thio-compounds, and
➢ Halide leaching.

Table 3 The sustainable evaluation of various leaching methods as a function of environmental, economic, and technological reliability [22, 23]

Lixiviant employed	Environmental impact	Kinetics	Economic feasibility			Technological reliability
			Consumption cost	Corrosiveness	Score (0–5)	
Cyanide	High	Slow	Cheaper	Non-corrosive	4.46	High
Aqua regia	High	Rapid	High	Corrosive	3.48	High
Thiosulfate	Non-toxic	Medium	High	Non-corrosive	2.71	Low
Thiourea	Low	Medium	High	Non-corrosive	4.00	Medium
Chlorine	Medium	Fast	Low (*in situ* generation and recycling)	Highly corrosive	3.25	Medium
Iodide	Non-toxic	Fast	High	Non-corrosive	3.64	Medium

2.3.1 *Leaching with sulfur-based thio-compounds*

For gold leaching, the useful sulfur-based thio-compounds are thiosulfate, thiourea, and thiocyanate [24]. Besides the complications in solution chemistry exhibited by the instability of thio-compounds and many other reactions which simultaneously take place during the gold leaching, these thio-compounds have chemical similarities to enable gold to be leached under an oxidizing environment.

Among them, thiocyanate [SCN]$^-$ is analogous to the cyanate ion [OCN]$^-$, where oxygen is replaced by sulfur [25]. The similarity in reactions with halides and being non-toxic in contrast to cyanide even in acidic condition are advantageous, however, it is not much studied in the case of MP-PCBs/e-waste. One of the strong reasons may be the presence of high amount of copper in such waste which can form insoluble Cu(I)–CN complex and causes faster degradation of thiocyanate solution [26], otherwise it requires very high oxidizing environment to keep the copper as Cu(II)-ions which also get precipitated as follows [27]:

$$Cu^{2+} + 2SCN^- = Cu(NCS)_2 = CuSCN + \frac{1}{2}(SCN)_2. \qquad (1)$$

Hence, only the processes and applications of thiosulfate and thiourea are being discussed here.

2.3.1.1 *Thiosulfate leaching*

An alternate to the cyanidation and/or aqua regia leach process, thiosulfate $(S_2O_3)^{2-}$ has been widely accepted as the best suitable reagent. Not only because the thiosulfate system is less threatened to the environment but also due to the achievable faster kinetics than that of the cyanide leaching [19,22–24]. The chemistry of gold–thiosulfate system is complex and needs an oxidizing atmosphere to keep reactions under control, for which copper is commonly being used as an oxidizing agent with self-catalytic behavior. The compulsion to maintain the alkali condition in thiosulfate system to prevent its decomposition by acid is fulfilled in ammoniacal medium; under such conditions copper can easily form the amine complexes to catalyze the reaction kinetics [28]. In the case of gold extraction from the e-waste/ MP-PCBs, the amount of copper therein with less interference of foreign metals due to its inability to form the soluble ammine complexes is another

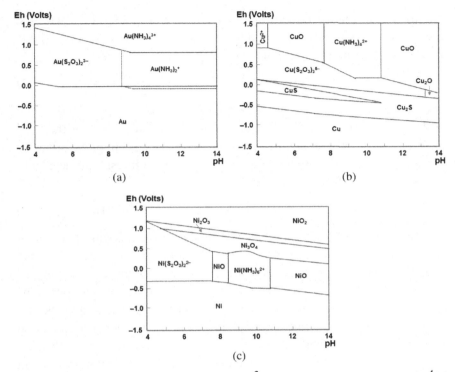

Fig. 5 The Eh–pH diagram for the (a) Au–$(S_2O_3)^{2-}$–H_2O system (conditions: 5×10^{-4} M Au, $1MS_2O_3^{2-}$, 1M NH_3/NH_4^+), the dotted line marks stability region of $Au(NH_3)_2^+$; (b) Cu–$(S_2O_3)^{2-}$–H_2O system (conditions: 0.5 M Cu^{2+}, $1MS_2O_3^{2-}$, 1M NH_3/NH_4^+); (c) Ni–$(S_2O_3)^{2-}$–H_2O system (conditions: 0.35 M Ni^{2+}, $1MS_2O_3^{2-}$, 1M NH_3/NH_4^+); all at 25°C

potential cause to be useful for the thiosulfate leaching. Gold alloyed with nickel followed by nickel and copper layers (as shown in Fig. 3(c)) can easily be liberated to leach in ammoniacal solution as all the three metals form their ammine complexes [29–31]. The potential-pH diagrams of Au, Cu, and Ni are given in Figs. 5(a)–5(c), respectively.

There are so many entities for gold leaching in the complex system like the simultaneous presence of complexing ligands thiosulfate and ammonia, the redox couple of Cu(I) and Cu(II) and the stability of thiosulfate itself under certain conditions of acid/alkali. The understanding of the aqueous chemistry is therefore of vital importance at first. The two gold–thiosulfate complexes are known to form as $Au(S_2O_3)^-$ and $Au(S_2O_3)_2^{3-}$ with the latter complex being the most stable [32]. The plausible reaction with O_2 used as the oxidant can be written as follows:

$$4Au + 8S_2O_3^{2-} + O_2 + 2H_2O = 4[Au(S_2O_3)_2]^{3-} + 4OH^-. \qquad (2)$$

But, the above reaction has been found to exhibit slow kinetics due to the passivation of sulfur that decomposed of thiosulfate on the gold surfaces [33]. Introducing the ammonia prevents such passivation by preferential adsorption onto gold surfaces over the thiosulfate, and gold can be leached as follows [29, 34]:

$$Au(NH_3)^{2+} + 2S_2O_3^{2-} = [Au(S_2O_3)_2]^{3-} + 2NH_3. \qquad (3)$$

However, the formation of gold–ammine complex is only possible at higher temperature [35], and $>80°C$ is economically not a good choice with high rate of ammonia decomposition at an elevated temperature ($>60°C$). This has been encountered by the catalytic action of copper ions in gold–thiosulfate–ammonia leaching [36]. The leaching of gold occurs under oxidizing environment of Cu(II) and Ni(II) has been found to be beneficial with an enhanced kinetics by 18–20 fold [30, 31, 37], that too at a temperature $<60°C$ [38]. Moreover, the gold–thiosulfate–ammonia leaching in the presence of copper and nickel is an electrochemical reaction, in which the $Cu(NH_3)_4^{2+}$ gets converted to $Cu(NH_3)_2^+$ to support the formation of oxidized product $[Au(S_2O_3)_2]^{3-}$; the reverse oxidation of Cu(I) to Cu(II) occurs in the presence of O_2 [39, 40]. The reactions that take place in this case are as follows:

$$Au + 2S_2O_3^{2-} = [Au(S_2O_3)_2]^{3-} + e^-, \qquad (4)$$

$$Cu(NH_3)_4^{2+} + 3S_2O_3^{2-} + e^- = [Cu(S_2O_3)_3]^{5-} + 4NH_3, \qquad (5)$$

$$2Cu(NH_3)_4^{2+} + 8S_2O_3^{2-} = 2[Cu(S_2O_3)_3]^{5-} + S_4O_6^{2-} + 8NH_3, \qquad (6)$$

$$2[Cu(S_2O_3)_3]^{5-} + 8NH_3 + \frac{1}{2}O_2 + H_2O$$

$$= 2Cu(NH_3)_4^{2+} + 8S_2O_3^{2-} + OH^-, \qquad (7)$$

whereas in contrast to the ore bodies, liberation of gold from the MP-PCBs is much dependent on how the alloyed nickel with gold layer is behaving in the ammonia–thiosulfate system. This makes the system much more complicated than the earlier complexed system of Au–Cu–NH_3/NH_4–S_2O_3 in solution, and hence has hardly been described. The thermodynamic stability of Cu(II)–thiosulfate complex is higher than $Cu(NH_3)_4^{2+}$; due to this, the $Cu(NH_3)_2^+$ gets reduced to a Cu(II)–thiosulfate complex, causing an oxidation of thiosulfate to degrade as tetrathionate. Such a condition does not arise in the presence of nickel which also controls the thiosulfate consumption during the entire leaching process of gold from the MP-PCBs. The gold leaching can be catalyzed by the nickelous oxide under the similar Eh–pH condition of ammonia–thiosulfate system (Fig. 5(c)). In order to provide the clear picture of the complicated electrochemical reactions and phenomenon, the plausible mechanism is presented in Fig. 6.

It can be seen that the thiosulfate ions react with Au(I) on the anodic surface of gold and enter the solution to form $Au(S_2O_3)_2^{3-}$ which is catalyzed by the Cu(II)–Cu(I) ammine complexes. The reduction of $Cu(NH_3)_4^{2+}$ transfers two ammonia ligands to form the kinetically favored diaminoaurate(I) complex, which subsequently exchanges the ligands

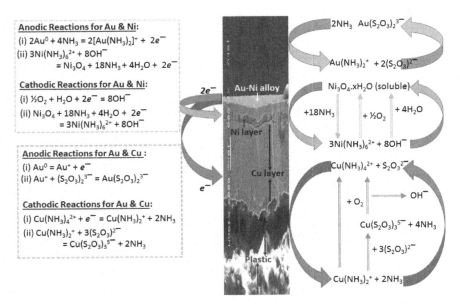

Fig. 6 The plausible electrochemical mechanism for the gold–thiosulfate leaching from MP-PCBs

with free thiosulfate ions to form the thermodynamically more stable aurothiosulfate complex. At the same time, the gold leaching is boosted by the oxidation reaction of Ni(II)–ammine complex to form Ni_3O_4, and then reduction of Ni_3O_4 with oxidation of gold as the $Au(NH_3)_2^+$ complex. The predominant cathodic reactions are dependent on relative concentrations of the species. Such a favorable environment in the presence of nickel is evident for the thiosulfate leaching of a typical silicate ore [41]. Moreover, plenty of works reported earlier for the $Au–Cu–NH_3–S_2O_3$ leaching process is diffusion controlled. But, leaching of gold from waste MP-PCBs (in which Au is alloyed with Ni, Fig. 3(c)) has been found to be chemically controlled, indicating the other entity that has influential involvement in gold leaching, and presence of nickel in the system as per the electrolytic mechanism narrated in Fig. 6 can be accountable for this. The involvement of nickel ammine complexes can also be understood from the EDX analysis of before and after leaching of waste MP-PCBs in ammonia-thiosulfate medium (Figs. 7(a) and 7(b), respectively) [42]. It can easily be depicted that along with gold, the entire nickel was leached in ammoniacal thiosulfate solution, and only after this the copper layer gets exposed to the lixiviant.

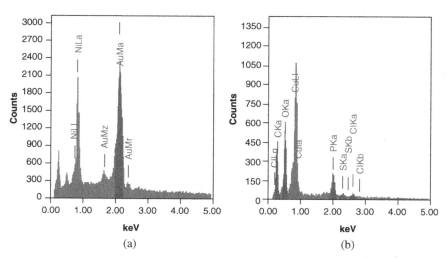

Fig. 7 The EDX analysis of (a) before and (b) after leaching of waste MP-PCBs in ammonia–thiosulfate medium [6]

However, a handful of research works directly dealing with the gold leaching from waste MP-PCBs in ammoniacal thiosulfate solutions have been studied; the role of influential factors is vital to understand.

• **Effect of thiosulfate concentration:** Using an optimal amount of thiosulfate in gold leaching is a necessity not only in terms of leaching efficacy but also to process cost control by lowering the reagent consumption, preventing the fast degradation and increasing the recycling possibilities. A variation in thiosulfate concentration caused change in the stability region of the gold species which also depends on the pH of the solution [43]. The leaching test of waste MP-PCBs (containing 0.12% Au, 35.1% Cu, 4% Sn, 2.7% Pb) with thiosulfate in a concentration range of 0.06–0.2 M showed a parabolic type leaching behavior of gold. Initially, the gold leaching increased with respect to increasing concentration of thiosulfate and then decreased at higher concentrations. Such phenomenon is similar to the one described elsewhere showing that a higher lixiviant concentration causes an increase in degradation products of thiosulfate in the form of sulfate, trithionate, tetrathionate, polythionates, etc. [39, 40], possibly as follows:

$$4S_2O_3^{2-} + O_2 + H_2O = 2S_4O_6^{2-} + 4OH^-, \qquad (8)$$

$$S_2O_3^{2-} + Cu^{2+} + 2OH^- = SO_4^{2-} + H_2O + CuS, \qquad (9)$$

$$2S_2O_3^{2-} + Cu^{2+} = S_4O_6^{2-} + 2Cu^+. \qquad (10)$$

Working with a granular sample of waste MP-PCBs containing 0.021% Au, 0.15 Ag, 56.68% Cu, 1.4% Sn, and Al has also revealed similar results (Fig. 8) [44]. The concentration effect of $(NH_4)_2S_2O_3$ varied in the range of 0.05–0.25 M (under the experimental condition of 40 mM $CuSO_4$, pH of the ammoniacal solution 10.0–10.5, stirring speed 250 rpm, temperature 25°C, solid–liquid ratio 60 g/L, duration 8 h) has revealed that gold extraction increases up to 0.1 M $(NH_4)_2S_2O_3$ and there after it decreases. The analysis of copper in leach liquor (Fig. 8) clearly indicates the genesis of such behavior. The copper concentration in leach liquor was the maximum 2.2 g at 0.1 M $(NH_4)_2S_2O_3$ which subsequently decreased to 1.3 g at 0.25 M $(NH_4)_2S_2O_3$. The residue obtained by the H_2SO_4–H_2O_2 leaching of waste MP-PCBs while leached in NH_3–S_2O_3–Cu system in the next stage has given a similar trend for gold leaching [45].

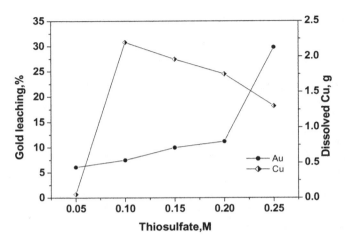

Fig. 8 Gold leaching efficiency and respective dissolution of copper as a function of ammonium thiosulfate concentration (experimental conditions: $CuSO_4$ 40 mM, pH 10.0–10.5, temperature 25°C, pulp density 60 g/L, time 8 h) [44]

The interactive relationship between thiosulfate and ammonia concentrations has shown significant effect on gold leaching rate from the MP-PCBs (Fig. 9) than the similar ligand-to-ligand interaction [42]. All these are supporting the formation of degradation products at higher concentration of thiosulfate. It is noteworthy to mention that the anionic degradation products like trithionate ($S_3O_6^{2-}$) and tetrathionate ($S_4O_6^{2-}$) do not have any lixiviating activity to further support the leaching anymore [46], and can interfere in gold recovery using the ion-exchange process by displacing the metal complexes from exchanger sites [47, 48]. An optimum ratio of ammonia to thiosulfate is therefore an essence to be maintained, at which copper can catalyze the gold leaching from the MP-PCBs without any formation of degradation products.

• **Effect of ammonia concentration and solution pH:** The role of ammonia in gold-thiosulfate leaching has been mainly identified for the stabilization of copper as the cupric ammine complex to catalyze the gold oxidation, which is helpful in further complexation with thiosulfate [49]. A higher ammonia concentration may cause the formation of solid species of copper like CuO, Cu_2O, and $(NH_4)_5Cu(S_2O_3)_3$ that creates hindrance on gold dissolution by getting coated on the gold surfaces. Such a behavior

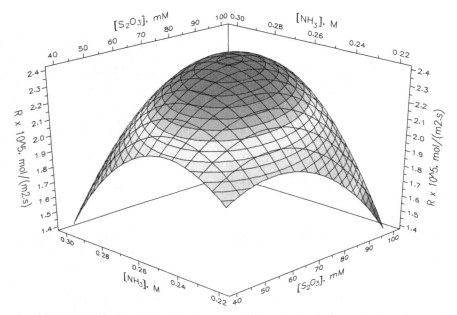

Fig. 9 Response surface analysis of the ammonia vs. thiosulfate on initial leaching rate of gold from MP-PCBs (at center level of cupric ions) [42]

has been confirmed by Ha *et al.* [43] while leaching the waste MP-PCBs with thiosulfate in the presence of 0.1–0.4 M ammonia in solution [43]. A higher leaching efficacy was obtained with 0.3 M ammonia at copper concentrations between 15 mM and 30 mM and $S_2O_3^{2-}$ concentration of 0.14 M, whereas lesser obtainment of gold in solution at lower ammonia concentration (<0.3 M) can be understood by the gold passivation with the decomposed products of thiosulfate at lower ammonia concentration. As it is clear from Fig. 5(a), gold leaching much depends on the thiosulfate concentration at solution pH < 9.0. For a similar kind of observation, the thiosulfate concentration in leach liquor of waste MP-PCBs as a function of pH (8.0–11.0) was analyzed [44]. The results indicated that an increase in pH up to 10.5 has given the highest solubility of thiosulfate ~0.1 M which decreased to 0.08 M at pH 11.0. Such behavior of thiosulfate as a function of pH matched well with the experimental results for gold leaching (presented in Fig. 10) and showed the maximum extraction at the solution pH 10.0–10.5.

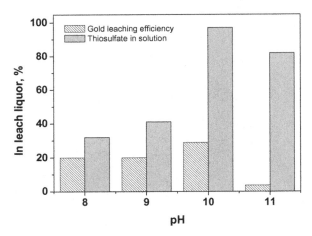

Fig. 10 Gold leaching efficiency as a function of pH and the corresponding thiosulfate analyzed to be soluble in the leach liquor at this particular pH (experimental conditions: 0.1 M $(S_2O_3)^{2-}$; 40 mM $CuSO_4$; 25°C temperature) [44]

The maximum rate of gold leaching was determined as 2.393×10^{-5} mol.m^{-2}.s^{-1} at 0.265 M NH$_3$ level [42]. It is imperative to mention that in the MP-PCBs, gold remains alloyed with nickel (Fig. 3(c)), therefore, the formation of nickel ammine complex also needs to be considered. The Eh–pH diagram shown in Fig. 5(c) diminishes the possibility for the formation of Ni(NH$_3$)$_6^{2+}$ at lower pH/ ammonia concentrations. This adversely affects the liberation of gold from Au–Ni alloy phase. Due to such an important role in gold leaching from MP-PCBs, the positive effect of NH$_3$ found among the terms analyzed is shown in Fig. 11.

• **Effect of copper concentration:** Maintaining the appropriate concentration in ammonia–thiosulfate system is an essential parameter to efficiently leach out the gold, however, gold leaching was negligible (<5%) without any addition of CuSO$_4$ in lixiviant solution and it increased with respect to increasing copper concentration at the maximum of 48 mM [44]. The results clearly attributed to the catalytic effect of copper in the gold leaching [50]. By varying the copper concentration in the range of 5–30 mM, it was found that going above 15 mM Cu has no further effect on gold leaching (Fig. 12) [43]. The gold leaching being independent to the Cu-concentration at above 15 mM can be corroborated with the switching over mechanism of gold leaching from diffusion control (at lower Cu-concentration) to

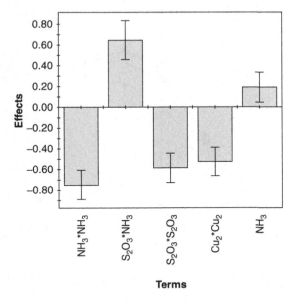

Fig. 11 The effects of terms on the initial rate of thiosulfate leaching of gold from MP-PCBs [42]

chemical control (at higher Cu-concentration) [51]. In addition to this, the stability region of copper species $Cu(NH_3)_4^{2+}$ and $Cu(S_2O_3)_3^{5-}$ is very much concentration dependent [28], and at very high Cu-concentration, precipitation of tenorite occurs. Therefore, in such a complex system of gold leaching, any excess of copper should be encountered by ammonia addition. Otherwise, it may cause formation of the thiosulfate oxidized product, tetrathionate [52], which could result in high reagent consumption with a decrease in gold leaching from the MP-PCBs.

• **Leaching kinetics of gold from the MP-PCBs:** The $Au–NH_3–Cu–S_2O_3$ system is very complex and reactions take place in several steps, mainly depending on the reagents concentration and nature of PCBs like size and composition. Ha *et al.* [43] have investigated the leaching kinetics for two different materials, waste mobile phone scraps and MP-PCBs at the optimized condition of 0.12 M thiosulfate, 0.2 M NH_3, and 20 nM Cu(II) [43]. The leaching kinetics for the mobile phone scrap was very fast and ~98% leaching completed within 2 h. In contrast, the leaching kinetics

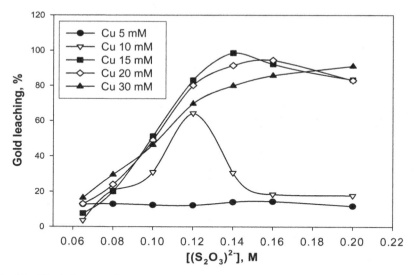

Fig. 12 The influence of copper concentration in gold leaching from waste MP-PCBs as a function of thiosulfate concentration in lixiviant [43]

for MP-PCBs was slower and reached only to 90% after 9 h of long duration. Moreover, the leaching rate drastically decreased ($\sim$20% leaching after 9 h) while 15 mM Cu(II) was used instead of 20 mM. Leaching of shredded MP-PCBs sample also showed lesser efficacy of gold leaching (30.3%) in comparison to 78.8% leaching of the complete MP-PCB unit under the same condition of 0.1 M $(NH_4)_2S_2O_3$, 40 mM $CuSO_4$, 40 g/L solid–liquid ratio and pH 10.0–10.5 [44].

The role of temperature in leaching kinetics of gold extraction from the MP-PCBs has been well established [42]. With constant parameters of 10 nM $CuSO_4$, 60 nM $Na_2S_2O_3$, 0.2 M NH_4Cl, and 0.26 M NH_3, leaching temperature varied in the range of 20–50°C. At the lowest temperature (20°C), the rate of gold leaching was at the lowest and only <50% gold could be extracted in 5 min, whereas at $\geq$40°C, the entire gold was leached within 2 min (Fig. 13). Such a difference in leaching kinetics as a function of temperature is corroborated with the reduction rate of Cu(II) to Cu(I). The rate of reduction for Cu(II) to Cu(I) at 50°C was 3.9-fold higher than that at 30°C (Fig. 13). The reaction rate at the reagent concentration range of 40–60 mM thiosulfate, 5–7 mM Cu(II) and 0.22–0.247 M ammonia was

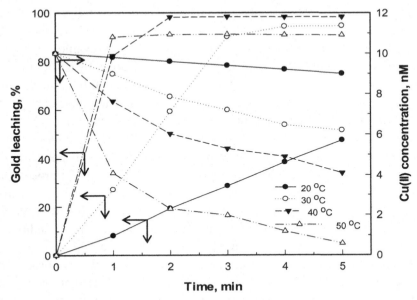

Fig. 13 The extraction kinetics of gold and reduction rate of Cu(II) to Cu(I) at different leaching temperatures of 20–50°C (experimental conditions: 10 mM CuSO$_4$; 60 mM Na$_2$S$_2$O$_3$; 0.2 M NH$_4$Cl; 0.26 M NH$_3$; 10 mL/min nitrogen sparging rate) [42]

given as follows:

$$R_1 = k[(S_2O_3^{2-})]^{0.40} \times [(Cu^{2+})]^{0.25} \times [(NH_3)]^{1.64}. \qquad (11)$$

While at the concentration range of 60–70 mM thiosulfate, 7–9 mM Cu(II), and 0.247–0.263 M ammonia, gold leaching rate is expressed as

$$R_2 = k[(S_2O_3^{2-})]^{0.22} \times [(Cu^{2+})]^{0.16} \times [(NH_3)]^{0.72}, \qquad (12)$$

those higher than the above-mentioned concentration range of the reagents can follow the zero-order kinetics [53].

- **Limitations and challenges ahead for the gold-thiosulfate leaching:** The very much concentration-sensitive thiosulfate process requires both dependent and independent optimization of each of the chemical components of the leach reaction. The thiosulfate leaching reactions are less favorable than the cyanide and aqua regia leach reaction for both the gold and silver recovery [54, 55], hence needed higher concentration to achieve equivalent rates of gold leaching. A typical thiosulfate leach solution will have a reagent concentration of 5–20 g/L vs. 0.25–1 g/L cyanide solution. The higher consumption of thiosulfate can partially offset by its significant lower cost as little as one fifth the cost of cyanide. It will be important to use the thiosulfate tailings solution either to recycle back in leaching and as the top-up volume or to productively use as a fertilizer.

In order to recover the gold values from the leach liquor, gold-thiosulphate complex is found to be poorly adsorbed by activated carbon, negating the use of conventional CIP or carbon-in-leach (CIL) processes. Instead, the evaluation of resin-in-pulp (RIP) technology to adsorb the gold from leach pulp onto strong-base anion exchanger should be more prominent because the RIP requires mild thiosulphate leaching conditions as the strong thiosulphate leach conditions may result in competitive adsorption on the resin sites by the anionic polythionate that are produced *in situ* during the leaching operation. Alternatively, the cementation of gold onto copper metal powder or some other reductant may be of potential downstreaming.

2.3.1.2 *Thiourea leaching*

Besides the ammonia-thiosulfate system, the use of thiourea with dilute mineral acids as a non-cyanide lixiviant for gold extraction has received much attention in recent years [19]. Despite being a suspected carcinogen [56], thiourea has certain advantages over the cyanide leaching. Thiourea is supposed to be low toxic reagent, exhibits faster gold leaching rate, alleged less interference from the base metals (Pb, Co, Ni, Zn), and moreover, it is used in acidic condition rather the alkaline condition of cyanide and thiosulfate [57]. The problems in gold leaching due to the occurrence of base metals in MP-PCBs and sorption loss of gold from leach solution with the carbonaceous matter coming from the detached hydrocarbon of the PCB resins can be handled in thiourea leaching.

The Eh–pH diagram for the $Au–SC(NH_2)_2–H_2O$ system is presented in Fig. 14. Thiourea is unstable and easily decomposed in neutral and alkaline pH range, therefore in general, the leaching is carried out in the pH range of 1–2 to obtain the only common cationic complex of gold,

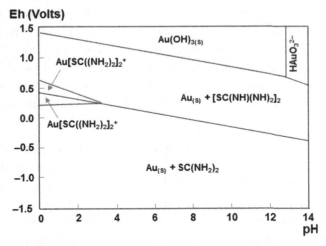

Fig. 14 The Eh–pH diagram of Au–SC(NH$_2$)$_2$–H$_2$O system at 25°C (conditions: 5 × 10^{-4} M Au, 0.5 M SC(NH$_2$)$_2$)

Au[SC(NH$_2$)$_2$]$_2^+$ [58]. Different oxidants, viz. sodium peroxide, hydrogen peroxide, ferric ions, oxygen, ozone, manganese dioxide, and dichromate are employed for the leaching, where ferric ions are the most effective in acidic sulfate solution than the chloride–nitrate solutions [59, 60]. The reaction for gold leaching in thiourea and ferric ion solutions can be written as follows:

$$Au + 2SC(NH_2)_2 + Fe^{3+} = Au[SC(NH_2)_2]_2^+ + Fe^{2+}. \qquad (13)$$

An oxidative degradation of thiourea into less stable formamidine disulfide [(NH$_2$)(NH)CS]$_2$ breaks down into cyanamide, thiourea, and elemental sulfur as follows:

$$2SC(NH_2)_2 + Fe^{3+} = [SCN_2H_3]_2 + 2Fe^{2+} + 2H^+, \qquad (14)$$

$$[SCN_2H_3]_2 = SC(NH_2)_2 + NH_2CN + S^0, \qquad (15)$$

$$SC(NH_2)_2 + Fe^{3+} + SO_4^{2-} = [FeSO_4.SC(NH_2)_2]^+. \qquad (16)$$

However, under mild acidic condition, thiourea may lead to hydrolysis decomposition as follows:

$$SC(NH_2)_2 + H_2O = (NH_2CONH_2) + H_2S. \qquad (17)$$

Both the reaction products of Eq. (17) are undesired species in gold leaching. Urea may decrease the leaching rate due to surface passivation, and H_2S may back-precipitate the gold from leach solution [19]. Therefore, the system of gold-thiourea leaching is also not simple to operate and the potential application at commercial scale is contingent upon the control on parameters like redox potential, pH, reagent concentration, time, etc. [14].

• **Effect of thiourea concentration:** As the aqueous chemistry described above for the $Au–SC(NH_2)_2–H_2O$ system, maintaining proper dosage of thiourea in gold leaching is primarily important. In general, for such a leaching system, 0.13 M thiourea along with the Fe(III) ion concentration up to 5 g/L at Eh and pH values nearly in the range of 400–450 mV and 1–2, respectively, is optimal [58]. In specific leaching of gold from the waste MP-PCBs, a few works have been reported. The pulverized sample of MP-PCBs containing 430 g/t Au, 540 g/t Ag, 39.86% Cu, 0.457% Zn, and 0.396% Ni has been leached in 0.26–0.36 M thiourea solution along with 0.6% Fe(III) at a pH 1.0 for 3 h duration [61]. At 0.26 M thiourea, the leaching rate was the slowest, however increased to the maximum (~90%) with 0.36 M thiourea in solution up to initial 2 h of leaching and then decreased significantly (to ~60% after 3 h) with prolonged leaching (Fig. 15). A concentration of

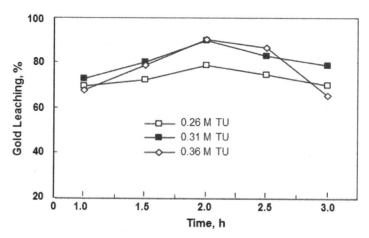

Fig. 15 Leaching kinetics of gold as a function of thiourea concentration with a constant Fe(III) concentration of 0.6% and pH value ~1.0 of the lixiviant [61]

0.31 M thiourea in lixiviant solution was optimized in this case to yield 89.7% Au along with 48.3% Ag for 2 h leaching time. In quite a similar leaching trend, the maximum leaching efficacy was with 0.31 M thiourea, below and above that concentration, it showed a remarkable decline in gold leaching [45]. Such behavior strongly supports the occurrence of passivation phenomena in leaching with the degradation products of thiourea as Eqs. (14–17). At a variance with these results, the optimized dosage of thiourea increased to 0.5 M when the calcined powder (size, 53–75 μm) of MP-PCBs (containing high amount of Au = 3470 g/t) was subjected to leaching [62]. At a pulp density of 2.85 g/L, the extraction efficiency was nearly 90% after 24 h of leaching with 0.5 M thiourea in 0.05 M acidic solution. Those lower than 0.5 M thiourea concentration could not exceed the gold extraction >50%, even at a lower pulp density than 2.85 g/L. Going above 0.05 M acid concentration could not benefit the leaching yield of gold.

• **Effect of ferric ions in thiourea solution:** Ferric ions are the most effective oxidants in gold-thiourea leaching in acidic media (pH 1–2), and the sulfate solution has been identified more suitably than the chloride/nitrate solutions [59, 60]. Gold leaching in thiourea solution with redox couple reactions of Fe(III)–Fe(II) has been found faster up to 4-fold than providing the oxidizing environment by air [63]. Therefore, the dosage of ferric ions is a very important parameter in gold-thiourea leaching.

Li *et al.* [61] have investigated the role of Fe(III) in the oxidation of gold and silver, and found that up to 0.6% ferric ions in the leaching solution increases the extraction of both metals. Lower dosage of ferric ions was not adequate for the oxidation of metals, while at higher dosage, thiourea also oxidized to S^{2-}, S^0, and formamidine disulfide which has reduced the recovery of gold and silver. At 0.6% mass concentration of Fe(III) in lixiviant solution, the leaching efficiency of gold reached the highest at 89.7% (Fig. 16) along with 48.3% silver extraction within 2 h duration. Working with the calcined fine powder of MP-PCBs [62], the maximum gold could leach in 2 h with an addition of 0.01 M ferric salt in solution under the condition of 2.85 g/L pulp density, 0.5 M thiourea, 0.05 M H_2SO_4. In the absence of any ferric salt, it took 6 h to achieve the same extraction of gold in acidic thiourea solution. However, when the ferric ion concentration

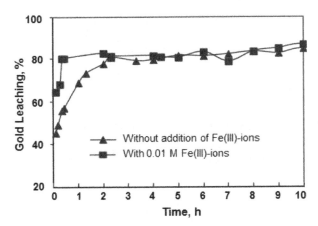

Fig. 16 Gold leaching kinetics with thiourea in the absence and presence of ferric ions (experimental conditions: 0.5 M thiourea in 0.05 M H_2SO_4; calcined MP-PCBs charged at pulp density 2.85 g/L; temperature 30°C) [62]

was higher than 0.01 M, gold recovery was lower due to the description in Eqs. (14–16). The iron present in PCBs may also supply a part of iron during the leaching. In the case of silver leaching, there was no beneficial effect of ferric ions. Since gold has lower oxidation potential values than silver, it seems reasonable that the presence of external oxidizing agent enhances the gold dissolution.

• **Effect of temperature on leaching mechanism:** In general, temperature always has a major role in the leaching kinetics and mechanism of the process. The calcine powder sample of MP-PCBs (3470 g/t Au) leached in a solution of 0.5 M thiourea + 0.05 M H_2SO_4 + 0.01 M Fe(III) ions at a pulp density 2.85 g/L with varying temperatures (30–60°C) has not shown much enhancement in gold extraction up to 45°C (with ~80% leaching), rather decreased remarkably (~75% leaching) at 60°C (Fig. 17) [62]. The waste MP-PCBs without calcination showed a similar trend in results [61]. The maximum 90% gold extraction efficiency was obtained at 25°C for 2 h leaching of a pulverized sample of 430 g/t Au with 0.31 M thiourea solution (with 6% Fe(III)) at pH 1. Such a behavior of gold leaching as a function of temperature and gold concentration in the charged sample can be elaborated with the complex electrochemical mechanism involved in the system.

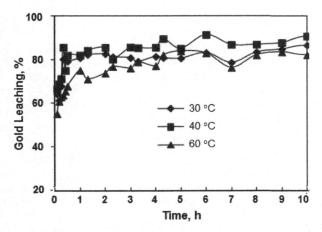

Fig. 17 The kinetics for gold extraction with thiourea as a function of leaching temperature (experimental conditions: 0.5 M thiourea + 0.01 M Fe(III) ions in 0.05 M H_2SO_4; calcined MP-PCBs charged at pulp density 2.85 g/L) [62]

The reason behind the suitability of higher temperature (up to 45°C) for Au-concentrated sample of calcined MP-PCBs in comparison to that (25°C) for the direct pulverized sample of lesser gold can be explained as an increase in gold concentration reduces the stability zone of $Au[SC(NH_2)_2]_2^+$ complex [64]. A decrease in gold extraction with a rise in temperature can be understood by the thiourea decomposition, which results in the formation of colloidal sulfur in the presence of iron (as Eqs. (14) and (15)) and retardation of reaction through passivation on the gold surfaces [65]. A significant effect of temperature on gold leaching with respect to increasing temperature can be corroborated with chemically controlled reaction [66], also evidenced by the independent leaching rate to the stirring speed. This is found at variance with the reaction kinetics investigated using rotating disc by applying formamidine disulfide as an oxidant instead of ferric ions. The chemically controlled gold leaching from MP-PCBs is expected to be an electrochemical reaction as schemed in Fig. 18.

The cathodic half-cell reaction (a) for the reduction of formamidine disulfide (in Fig. 18) seems like the H^+-ion might participate in the rate-limiting reaction. But, it has been found that the pH does not change with decomposition of thiourea, indicating that the formamidine disulfide exists in a protonated form (as reaction-b) instead of a neutral molecule [67, 68].

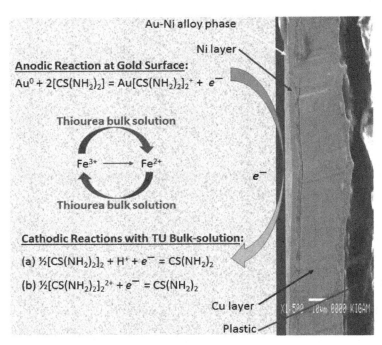

Fig. 18 The plausible electrochemical mechanism for the gold-thiourea leaching from MP-PCBs

The presence of Fe(III)-ions in the bulk thiosulfate solution catalyzes the electrochemical reactions to facilitate the oxidation of metallic gold to aurous (Au^+) and silver to argentous (Ag^+) ions, as shown in Fig. 18 (for Au^0 to Au^+).

• **Limitations and challenges ahead for the gold-thiourea leaching:** More or less, the challenges for the thiourea leaching process of gold are the same as using the thiosulfate. Without an oxidizing agent, thiourea solution alone is less effective. $Fe_2(SO_4)_3$ is preferred to H_2O_2 since it increases gold extraction significantly, however, it does increase thiourea consumption. It should also be noted that an excess of $Fe_2(SO_4)_3$ does not favor gold leaching and falls down rapidly. The initial concentration of thiourea with respect to $Fe_2(SO_4)_3$ concentration and the oxidation kinetics of thiourea are the most important factors affecting the reaction chemistry of gold in at pH 1–2. The control of thiourea leaching system is very difficult due to its oxidation in successive stages to a number of degradation products.

Considerably high consumptions of reagents like $SC(NH_2)_2$, $Fe_2(SO_4)_3$, and H_2SO_4 make the process expensive in comparison to cyanidation. The gold with Ni and on the surface of Cu-layer in MP-PCBs is also contributing in the reaction to form an insoluble complex with thiourea and the thiourea degraded sulfur, which further increases the thiourea consumption. It may be due to this reason that Camelino *et al.* [45] found the thiourea leaching efficiency of gold to be much lesser (45%) than the thiosulfate leaching ($\sim$70%). An oxidative leaching step in sulfuric acid, prior to the thiourea leaching, is therefore suggested by some researchers that can preliminarily remove the base metals (M, as denoted in Eq. (18)) and thereafter the remaining values of precious metals can be leached in thiourea solution as described above [69,70]. Hence, without carefully handling the complexity, it would be difficult to implement the thiourea leaching in larger operation of gold recovery through recycling of waste MP-PCBs.

$$M^0 + H_2O_2 + H_2SO_4 = M^{2+} + SO_4^{2-} + 2H_2O. \tag{18}$$

2.3.2 *Leaching with halides*

Most of the halogens (except fluorine and astatine) have been tested for gold extraction due to its ease in complexation and exhibition of faster leaching rate other than using the aqua regia [71]. The highest leaching rate in aqua regia solution can be understood by the strong oxidizing environment provided to form soluble Au(III) species as the following reactions [72]:

$$3HCl + HNO_3 = NOCl + Cl_2 + 2H_2O, \tag{19}$$

$$Au + NO_3^- + 4H^+ = Au^{3+} + 2H_2O + NO, \tag{20}$$

$$Au^{3+} + 4Cl^- = AuCl_4^-, \tag{21}$$

$$Au + HNO_3 + 4HCl = HAuCl_4 + 2H_2O + NO, \tag{22}$$

where the various species of chlorine, Cl_2, Cl^-, and NOCl play an important role. In any halide media (of Chlorine/bromine/iodine) depending on the conditions of solution, however, gold forms both the Au(I) and Au(III) complexes with chlorine, bromine, and iodine. Numerous studies have been done for gold extraction from ore bodies using the halides, but a few directly deals with the application in recycling of gold from the waste MP-PCBs.

2.3.2.1 *Chlorination leaching*

Chlorination was well in practice earlier to the inception of gold cyanidation. The advantageous use of chlorination has been observed in the elimination of *in situ* adsorption loss of gold with carbonaceous matter, gold refining and cyanide detoxification processes [10]. Nevertheless, the requirement of special construction materials and handling of chlorine gas to withstand the severe condition are now being handled by using the best engineering design and *in situ* generation of gaseous chlorine. *In situ* chlorine generation is mainly done by employing two methodologies: (i) reactions of sodium hypochlorite by mixing with hydrochloric acid, and (ii) anodic generation of gaseous chlorine from an HCl solution in a designed electrolytic cell. With sodium hypochlorite system, the reaction to chlorine generation is supposed to take place as follows:

$$NaOCl + NaCl + 2HCl = 2NaCl + Cl_{2(g)} + H_2O, \tag{23}$$

whereas, in an electrolytic cell, gaseous chlorine can be generated at the anodic electrode surface by passing a certain amount of current as per the reaction:

$$\text{At anode:} \qquad 2Cl^- = Cl_{2(g)} + 2e^-, \tag{24}$$

$$\text{At cathode:} \qquad 2H^+ + 2e^- = H_{2(g)}. \tag{25}$$

The gaseous chlorine has high solubility in acidic water as follows [73]:

$$Cl_{2(g)} = Cl_{2(aq)}. \tag{26}$$

At a pH above 2.5, the aqueous chlorine predominantly forms HOCl as follows [10]:

$$Cl_{2(aq)} + H_2O = HCl + HOCl. \tag{27}$$

Both the acids, HCl and HOCl, completely dissociate in aqueous solutions as follows:

$$HCl = H^+ + Cl^-, \tag{28}$$

$$HOCl = H^+ + OCl^-. \tag{29}$$

Further, the soluble chlorine can react with the chloride ions to form the trichloride ions under high acidic condition (<pH 3):

$$Cl_{2(aq)} + Cl^- = Cl_3^-. \tag{30}$$

With dissolved species of chlorine in aqueous solution, gold can form soluble chloride complexes as follows [74]:

$$Au + Cl^- + \frac{3}{2}(Cl_2)_{aq} = AuCl_4^-, \tag{31}$$

$$Au + \frac{3}{2}HOCl + \frac{3}{2}H^+ + \frac{5}{2}Cl^- = AuCl_4^- + \frac{3}{2}H_2O. \tag{32}$$

The formation of gold chloro-complex as per Eqs. (31) and (32) also seems to be plausible with the Eh–pH diagram (Fig. 19) at pH < 6 [6]. It is very interesting to know about how the kinetics for the formation of gold chloro-complex differs with respect to the chlorine species. As shown in Eqs. (31) and (32), there are direct formations of auric chloride ($AuCl_4^-$) with the $Cl_{2(aq)}$ and HOCl, which are at variance with the reaction of less oxidative Cl^- ions. In the presence of the Cl^- ions, gold leaching mechanism takes place in several steps with the formation of an intermediate

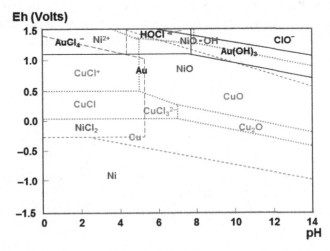

Fig. 19 The Eh–pH diagram of Au–Cu–Ni–Cl–H$_2$O system at 25°C (conditions: 5 × 10^{-4} M Au; 0.5M Cu^{2+}; 0.35 M Ni^{2+}; 2 M Cl$^-$)

aurous complex on gold surface as follows [75, 76]:

$$2Au + 2Cl^- = 2AuCl^-. \tag{33}$$

Due to the requirement for complex stability and the hard-and-soft acid base (HSAB) principle, high electronegative 'hard' donor atoms of chloride ions lead to form the metal complexes of high valance [77]. This indicates that the chloro-complex of monovalent gold with hard donor ligands will probably be disproportionate to trivalent and some of zero-valent gold according to the following reaction in the second step:

$$2AuCl^- = AuCl_2^- + Au. \tag{34}$$

Subsequently, the secondary intermediate compound $AuCl_2^-$ either undergoes to be oxidized to stable auric complex as follows [75, 76]:

$$AuCl_2^- + 2Cl^- = AuCl_4^-. \tag{35}$$

As shown in Fig. 3(c), the leaching of nickel that is alloyed with gold in the outer layer of MP-PCBs also takes place in this case. The diffusion of the dissolved chlorine, $Cl_{2(aq)}$ and Cl^- through a Nernst boundary layer from the bulk solution to the surface of Au–Ni alloy of MP-PCBs is, therefore in all possibility, the rate determining step for the leaching reaction of the both gold and nickel in dissolved chlorine solution. The mechanism of leaching for gold takes place as above reactions, whereas nickel gets leached out in the chlorine leaching as follows [78, 79]:

$$Ni + Cl_{2(aq)} = Ni^{2+} + 2Cl^-, \tag{36}$$

$$Ni^{2+} + Cl^- = NiCl^+, \tag{37}$$

$$Ni^{2+} + 2Cl^- = NiCl_2. \tag{38}$$

Similar to the nickel, the much higher copper content (wt%) there in MP-PCBs can also be leached in aqueous solution of $Cl_{2(aq)}$ and HOCl with thermodynamic feasibility as follows [80]:

$$Cu + Cl_{2(aq)} = Cu^{2+} + 2Cl^-, \tag{39}$$

$$Cu^{2+} + Cu + 2Cl^- = 2CuCl. \tag{40}$$

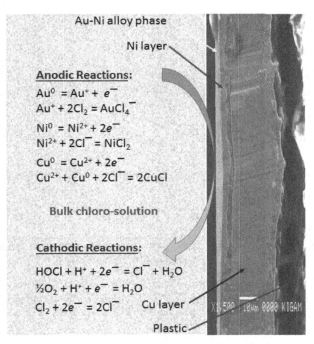

Au-Ni alloy phase

Ni layer

Anodic Reactions:

$Au^0 = Au^+ + e^-$

$Au^+ + 2Cl_2 = AuCl_4^-$

$Ni^0 = Ni^{2+} + 2e^-$

$Ni^{2+} + 2Cl^- = NiCl_2$

$Cu^0 = Cu^{2+} + 2e^-$

$Cu^{2+} + Cu^0 + 2Cl^- = 2CuCl$

Bulk chloro-solution

Cathodic Reactions:

$HOCl + H^+ + 2e^- = Cl^- + H_2O$

$\frac{1}{2}O_2 + H^+ + e^- = H_2O$

$Cl_2 + 2e^- = 2Cl^-$ **Cu layer**

Plastic

Fig. 20 The plausible electrochemical mechanism for the chlorine leaching of gold from MP-PCBs

Based on the above reactions, the plausible electrochemical mechanism for the chloro-leaching of gold (along with Ni and Cu that remains present in MP-PCBs) is presented in Fig. 20. The dissolution of large amount of base metals in the chloro-leaching has been found to be problematic. The comparative lower concentration of gold in leach liquor can present difficulty in subsequent downstream processing. The best answer came via the solution chemistry of gold [6]. At an ORP value <350 mV, the maximum of base metals from the MP-PCBs can be leached out in an acidic chloride solution by leaving gold in the residues. The Eh–pH diagram for an Au–Cu–Ni–Cl–H_2O system (shown in Fig. 19) also indicates to the plausible selectivity in leaching, as the copper and nickel forms soluble species at lower potential than the required potential to form the chloro-complexes of gold.

• **Effect of HCl concentration:** Interestingly, the increasing solubility with respect to an increase in HCl solution shows different behavior of

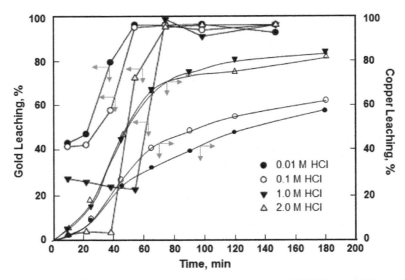

Fig. 21 The leaching rate of gold and copper from waste MP-PCBs at different HCl concentrations (experimental conditions: 2.0 M Cl$^-$ concentration; electrogeneration rate of Cl$_2$ at 714 A/m^2 current density; pulp density 17 g/L; temperature 25°C; particle size $-3/+2$ mm) [6]

gold leaching at lower (up to 0.1 M) and higher ($>$0.1–2.0 M) acidic condition (Fig. 21), however, copper leaching is consistently found to increase. The formation of different chlorine species as a function of acid concentration is mainly responsible for this behavior [6]. The HOCl formed at lower acidic condition has higher oxidizing power towards gold and that is evidenced by more favorable thermodynamic value than with Cl$_{2(aq)}$ ($\Delta G^\circ_{298\,K} = -113.2$ kJ and -139.4 kJ for Eqs. (31) and (32), respectively). In other way, it can be better defined with the leaching rate and mechanism because the thermodynamic data never give the rate of reactions. Figure 21 depicts that gold extraction shoots up after 45 min of leaching at the same condition of 2.0 M HCl. The highly acidic environment attacks copper from the boundary side layer, making a path enter inside the layer gap with higher leaching rate of copper. Subsequently, the leaching of gold from Au–Ni alloy phase gets started with increase in the soluble chlorine concentration in lixiviant after a prolonged feeding. The rate of reaction can also be enhanced by dissolved Cu(II) ions (as stated in Eq. (39)). Therefore, after 90 min of leaching, the efficacy of gold extraction was obtained at the maximum level (>99%).

• **Effect of temperature:** In general, an increase in chlorine mass transfer rate by elevation of temperature should support enhancement of the gold leaching efficiency [75, 81]. But in contrast, the solubility of chlorine (as $Cl_{2(aq)}$) decreases with increase in the temperature (from 7×10^{-3} M Cl_2 to 2.5×10^{-3} M Cl_2 at 20–60°C temperatures, respectively) due to the lesser absorbability of gases at higher temperatures. In chlorine leaching of waste MP-PCBs, therefore, copper extraction is found to decrease with increase in temperature [6]. At the initial stage of gold leaching, the lowest concentration of residual chlorine was mainly responsible for slower kinetics for the reactions of Eqs. (31) and (32) which accelerated after 40 min and after a 2 h prolonged leaching the gold extraction is found to be independent of the effect of temperature (Fig. 22). The favorable chemical kinetics with diffusion rate of hypochlorous species at higher temperature

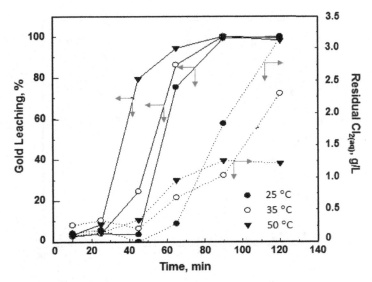

Fig. 22 Gold leaching rate at different temperatures and corresponding residual chlorine in HCl leached solution (experimental conditions: 2.0 M HCl solution; electrogeneration rate of Cl_2 at 714 A/m^2 current density; pulp density 17 g/L; temperature 25°C; particle size −3/+2 mm) [6]

can be considered for such behavior. The diffusion coefficients of $Cl_{2(aq)}$, HOCl and OCl$^-$ also confirm the suitability of temperature at 50°C [82].

• **Effect of initial Cl$^-$ ions concentration:** The dissolution kinetics of gold in chloride solutions has revealed the dissolution rate to be proportional to the chlorine–chloride concentrations [75]. Hence, the gold extraction from waste MP-PCBs is found to increase with increase in the initial concentration of adsorbed/soluble chlorine in the lixiviant of 2 M HCl [6]. The presence of chloride solution instead of water also accelerates the chlorine dissolution rate due to the retarding effect of Cl$^-$ ions on chlorine dissolution [83]. Therefore, the amount of initial chlorine in lixiviant enhances the leaching kinetics of gold by shifting the reaction from diffusion control to chemical control at a temperature of 25°C (Fig. 23). An increase in initial concentration of chlorine in the lixiviant did not much affect the efficacy of copper extraction as a function of time. It showed the plausible selectivity as a function of initial chlorine concentration within the leaching duration of 45 min, and established the parameter as an important one to be considered in selective gold leaching.

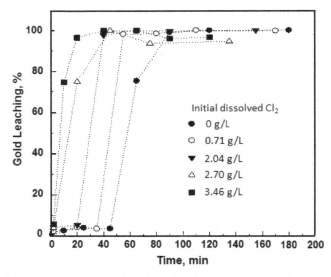

Fig. 23 Gold leaching rate as a function of initial concentration of dissolved chlorine in lixiviant, 2 M HCl solution (experimental conditions: electrogeneration rate of Cl$_2$ at 714 A/m^2 current density; pulp density 17 g/L; temperature 25°C; particle size $-3/+2$ mm) [6]

Table 4 The conditions and results of a 2-stage chlorine leaching (under controlled ORP) of waste MP-PCBs in HCl solution [6]

Leaching stages	ORP$_{(Ag/AgCl)}$	At optimal conditions	Metals in LL (mg/L)		% Extraction of metals	
			Au	Cu	Au	Cu
1st stage	<350 mV	40 g/L PD; 714 A/m^2 CD; 2 M HCl; 165 min	0.9	24000	5.0	94.9
2nd stage	>1100 mV	160 g/L PD; 714 A/m^2 CD; 0.1 M HCl; 10 min	67	0.6	93.1	0.6

PD = pulp density; CD = current density; LL = leach liquor.

- **Effect of the ORP on gold leaching:** As discussed with Eh–pH diagram of Au–Cu–Ni–Cl–H$_2$O system, the plausible selectivity in leaching of base metals over gold is observed. Kim *et al.* [6] has investigated for such selectivity under controlled ORP system. In the first stage of leaching at 350 mV ORP (other conditions given in Table 4), the maximum of copper leaching (95%) could be obtained along with the minutely extracted gold (0.9 mg/L) in leach liquor (Table 4). In the subsequent step, the residues of the first stage of leaching has yielded 93% gold with only 0.6 mg/L copper in the second stage of leaching performed at >1100 mV ORP (other conditions in Table 4). The leach liquor obtained has been found to be suitable for downstream processing via adsorption onto ion exchange resin, Amberlite XAD-7HP. With the maximum adsorption capacity of 46.03 mg Au/g resin has given concentrated eluted solution of 6 g/L gold with 99.9% purity is identified as a process of interest for gold recovery from the waste MP-PCBs.

- **Limitations and challenges for the chlorine leaching:** Though the highly corrosive acidic environment can be tackled by special stainless steel and rubber lined equipment and difficulties in transportation can be handled by *in situ* chlorine generation, the poisonous nature of chlorine is still a problem to be controlled to avoid any health risk. The use of chlorine with safety measurements as successfully since 1971 in Nevada for commercial operations [10], and in leaching of precious metals recently employed by Sumitomo Metal Mining Co [84] are attractive to be employed in large scale to recover the gold back from waste MP-PCBs. The obtainment of selectivity other than the controlled ORP leaching seems to be difficult, and without that in highly oxidizing environment of acidic media, the maximum of base dissolved in leach liquor can present big challenges during downstream processing. Moreover, control of ORP at production level would be challenging as Cu(II) itself is able to provide favorable oxidizing environment for gold leaching. Temperature plays an important role in chlorine adsorption in HCl solution and hence required controlled temperature reactions to lower the operational cost. Adsorptive dissolution of chlorine has already been identified as playing a major role in the chlorine leaching; therefore, it would be interesting to look at the replacement of HCl with other cheap and comparatively less corrosive reagents like NaCl.

2.3.2.2 *Iodine leaching*

Of the halogens, the gold–iodide complexes are the most stable in aqueous solutions. As the stability of the both Au(I) and Au(III) tends to decrease as the tendency of an atom to attract the electrons increases with increase in electronegativity of the ligand donor atoms [76]. This leads to the stability order of $I^- > Br^- > Cl^-$ for gold complexation. The Eh–pH diagram of the $Au–I_2–H_2O$ system (Fig. 24) demonstrates the stability of AuI_2^- complex

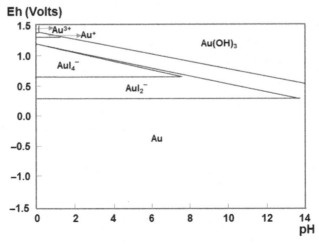

Fig. 24   The Eh–pH diagram of Au–I–H_2O system at 25°C (conditions: 5×10^{-4} M Au; 0.1 M iodine)

which occupies a wider solubility range approximately upto pH 13 [85,86]. Gold oxidizes in the presence of iodide to form AuI_2^- at a potential of 0.51 V and above 0.69 V, the AuI_4^- complex becomes stable. The overall stability region of AuI_4^- is found to be greater than that of either $AuCl_4^-$ or $AuBr_4^-$. The solubility of solid iodine as $I_{2(aq)}$ in pure water (1.85×10^{-3} M at 25°C) occurs as follows [87]:

$$I_{2(s)} = I_{2(aq)}. \tag{41}$$

Further increase in I^- concentration greatly increases overall solubility of iodine by the formation of tri-iodide ions, I_3^- at the pH < 9, as follows [88]:

$$I_{2(aq)} + I^- = I_3^-. \tag{42}$$

Above the solution pH > 9, a hypoiodite and iodite ion forms as follows [88]:

$$I_3^- + H_2O = IO^- + 2I^- + 2H^+, \tag{43}$$

$$3I_3^- + 3H_2O = IO_3^- + 8I^- + 6H^+. \tag{44}$$

The tri-iodide ion remains present when solutions contain $>1.0 \times 10^{-3}$ M dissolved iodine [87]; consequently, this can serve as an oxidant for gold leaching according to the electrochemical reactions shown in Fig. 25, and overall reaction as below [86]:

$$2Au + I_3^- + I^- = 2AuI_2^-, \tag{45}$$

$$2Au + 3I_3^- = 2AuI_4^- + I^-. \tag{46}$$

The iodo-complex, AuI_4^- formed as per Eq. (46) is not stable because the AuI_4^- easily oxidizes iodide ion to iodine [19], and is itself reduced to gold(I) as AuI_2^- (as per the electrochemical leaching mechanism shown in Fig. 25). Gold leaching in iodine–iodide solutions has been mainly studied by electrochemical techniques as the leaching is not remarkably affected by changes in pH over a wide range. Therefore, the Nernst equation for the $[I_3^-]/[I^-]$ couple has been determined as follows [85]:

$$E = 0.54 - 0.03 \log \frac{[I^-]^3}{[I_3^-]}. \tag{47}$$

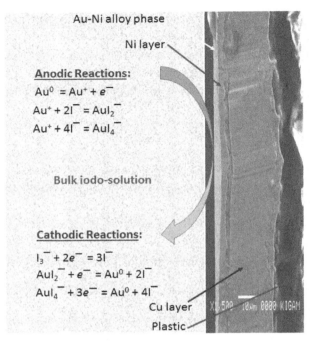

Fig. 25   The plausible electrochemical mechanism for the iodide leaching of gold from MP-PCBs

At the concentration of $[I^-]$ and $[I_3^-]$ equal to 1.0×10^{-3} M, the calculated 0.71 V of oxidation potential is sufficient enough to oxidize gold to form the complexes with iodine species (as above reactions) [76]. The direct application of iodide–iodine leaching of gold from secondary resources, such as MP-PCBs, is limited in the literature [89, 90] to study the technical feasibility under the different conditions.

• **Effect of oxidant:** The iodide leaching of gold in the presence of oxidant is advantageous as it increases the recovery and reduces the iodine consumption, thus improving the economics of the process. The leaching of gold from the fine particles of MP-PCBs is done using the hydrogen peroxide as oxidant in iodine solution [89, 90]. Addition of 1% H_2O_2 to the lixiviant solution has caused increase in the leaching yield of gold to 95%. A further addition of H_2O_2 has led to precipitating the iodine onto gold surface, resulting in declination of gold recovery. The reaction in the presence of hydrogen peroxide takes place as follows:

$$2Au + 4I^- + H_2O_2 = 2AuI_2^- + 2OH^-. \qquad (48)$$

Using two stages leaching of the waste MP-PCBs in KI–H_2SO_4–H_2O_2 media, the recovery of gold was 93.4% [91]. The maximum of copper (95.3%) in leach liquor has presented the application of acidic solution to be disadvantageous for gold leaching, and certainly for the high reagent consumption.

• **Effect of pretreatment on iodide leaching of gold:** To avoid the above-mentioned situation, a pretreatment with supercritical water oxidation (SCWO) followed by HCl leaching has been reported [92]. The comminuted waste MP-PCBs (<4 mm sized of composition: 65 g/t Au, 1060 g/t Ag, 50 g/t Pd and 40.8% Cu, etc.) pretreated under an elevated temperature of 370–450°C in the presence of H_2O_2 could leach the maximum of copper and other base metals in 1 M HCl solution (Fig. 26). The remaining solid leached with an iodide–iodide solution (varied in the range of 0–60 mM iodine; 50–275 mM KI) yielded the complete leaching of Ag in 90 min

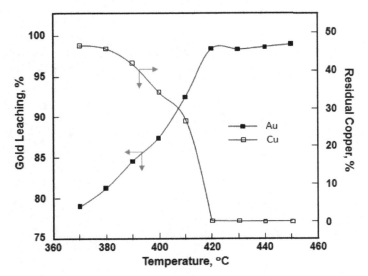

Fig. 26 The leaching efficacy of gold in iodine solution as function of SCWO temperature and corresponding residual copper in leached residues (experimental conditions: iodine/iodide mole ratio 1:5; pulp density 10%; pH 9; duration 120 min) [92]

and in next 30 min of leaching time, the entire Au and Pd get extracted. At a higher iodine-to-iodide mole ratio than 1:5, the formation of I_3^- is reduced causing the precipitation of AuI, AgI, and PdI_2, and decreased the leaching efficacy of Au, Ag, and Pd. On the contrary, high I^- concentration also causes decrease in the leaching efficacy of Au, Ag, and Pd due to the formation of CuI and PbI_2. The highest gold leaching efficiency of 98.5% was achieved at a pulp density of 10 g/L.

• **Effect of pH and iodine concentration:** The iodide leaching of gold and the associated silver and palladium from the residues of pretreated MP-PCBs is found to increase with increase in pH and achieved the maximum at 9 pH. Further increase in the pH of lixiviant has adversely affected the extraction efficiency (Fig. 27). On that condition, the IO_3^- became the predominant species in aqueous solution instead of I_3^- [88]. With a decreased concentration of active iodine, therefore, the AuI_2^-, AgI_2^- and PdI_3^- complexes became unstable and caused suppression of leaching of the gold, silver, and palladium. In addition, the hydroxide precipitates of other

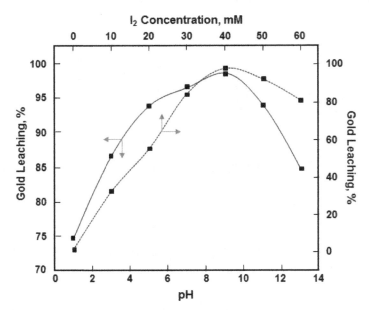

Fig. 27 The leaching efficacy of gold in iodine solution as function of pH and I_2 concentration in the lixiviant (SCWO conditions: temperature 420°C; acid 1 M HCl; duration 60 min) [92]

base metals there in waste MP-PCBs may form on the surface of precious metals under the strong alkaline system, leading to the decrease in gold leaching (with Ag and Pd). The maximum gold leaching was obtained with the iodine/iodide mole ratio at 1:5 (at I_2 concentration of 40 mM, as shown in Fig. 27). The excessive I^- reacts with residual base metals like CuO and Pb_2O to precipitate as CuI and PbI_2, respectively, that can passivate on the surface of gold and other precious metals to adversely affect their leaching. Under strong acidic environment, the formation of solid species (like AuI) can also passivate the surface of precious metals and retard the formation of complexes such as AuI_2^- to hinder the leaching efficiencies of precious metals.

• **Limitations and challenges for the iodide leaching:** Iodine has been identified as a good oxidant (<11 pH) and better than HOCl with higher leaching kinetics of gold [93]. Nevertheless, iodine continues to be under-employed as a lixiviant for gold leaching, may be largely because of cost [15, 57]. Besides this, the non-toxic and non-corrosive nature of iodine and the stability of gold–iodide complex at a wide range of pH can be advantageous if the research community takes it as a "Green" replacement of highly toxic cyanidation process. The potential prospect may be similar to the way presented by Kim *et al.* [6] for the selective chlorine leaching of gold in two stages (by controlling the potential). The controlled chlorine leaching in the forefront followed by iodide leaching at back-end may be an alternative, subject to process costing and compatibility of the two medium at two different stages.

References

[1] African Countries Mining & Mineral Industry Business Opportunities Handbook, Volume 1. West Africa (2014) 4th edition, International Business Publication, USA.

[2] World Demand Trends, Full Year (2014) A report by World Gold Council.

[3] Tuncuk A, Stazi V, Akcil A, Yazici EY, Deveci H (2012) Aqueous metal recovery techniques from e-scrap: Hydrometallurgy in recycling. *Minerals Engineering*, **25**(1): 28–37.

[4] Christian Hagelüken (2006) Improving metal returns and eco-efficiency in electronics recycling — a holistic approach for interface optimisation between pre-processing and integrated metals smelting and refining. In: Proceedings of

the 2006 IEEE International Symposium on Electronics & the Environment, 8–11 May 2006, San Francisco, pp. 218–223.

[5] Jeong J, Lee JC, Choi JC(2015) Characterization of metal composition in spent printed circuit boards of mobile phones. *Journal of the Korean Institute of Resources Recycling*, **24**(3): 76–80.

[6] Kim EY, Kim MS, Lee JC, Pandey BD (2011) Selective recovery of gold from waste mobile phone PCBs by hydrometallurgical process. *Journal of Hazardous Materials*, **198**: 206–215.

[7] Schluep M, *et al.* (2009) UNEP Sustainable Innovation and Technology Transfer Industrial Sector Studies. Recycling from E-waste to Resources. (Eds.) G. Sonnemann and B. de Leeuw. www.unep.org. Accessed 15 Sept. 2015.

[8] Meskers CEM, Hagelüken C, Damme GV (2009) Green recycling of EEE: Special and precious metal recovery from EEE. In: Proceedings of EPD Congress 2009, TMS 2009, (Eds.) S. M. Howard, P. Anyalebechi and L. Zhang, California, USA, pp. 1131–1136.

[9] Lee JC, Song HT, Yoo JM (2007) Present status of the recycling of waste electrical and electronic equipment in Korea. *Resources Conservation and Recycling*, **50**: 380–397.

[10] Marsden JO, House CL (2006) The Chemistry of Gold Extraction. 2nd edition, Society of Mining, Metallurgy, and Exploration, Inc. (SME) Colorado, USA.

[11] Koyama K, Tanaka M, Lee JC (2006) Copper leaching behavior from waste printed circuit board in ammoniacal alkaline solution. *Materials Transactions JIM*, **47**(7): 1788–1792.

[12] Habashi F (2005) A short history of hydrometallurgy. *Hydrometallurgy*, **79**(1–2): 15–22.

[13] Loconto PR (2015) A decade of quantitating cyanide in aqueous and blood matrices using automated cryotrapping isotopic dilution static headspace GC-MS. *LCGC North America*, **33**(7): 490–501. http://www.chromatography online.com/decade-quantitating-cyanide-aqueous-and-blood-matrices-using-automated-cryotrapping-isotopic-dilutio (Accessed 25 Sept. 2015).

[14] Ilyas S, Lee JC (2014) Biometallurgical recovery of metals from waste electrical and electronic equipment: a review. *Chem Bio Eng Review*, **1**: 1–23.

[15] Syed S (2012) Recovery of gold from secondary sources — a review. *Hydrometallurgy*, **115–116**: 30–51.

[16] Demopoulos GP, Cheng TC (2004) A case study of CIP tails slurry treatment: comparison of cyanide recovery to cyanide destruction. *The European Journal of Miner Processing and Environment Protection*, **4** (1): 1–9.

[17] Brug JE, Heidelberg EX (1974) Recovery of gold from solution in aqua regia. US Patent 3,856,507 A.

[18] Jeong J, Bae M, Srivastava RR, Jun M, Lee J-c, Shin Dy, Kim S-k (2015) Recovery of nitric acid and gold from aqua regia solution by solvent extraction. Korean Patent application no. 10-2015-0137378 (dated 9th September 2015).

[19] Sparrow GJ, Woodcock JT (1995) Cyanide and other lixiviant leaching systems for gold with some practical applications. *Mineral Processing and Extractive Metallurgy Review*, **14**: 193–247.

[20] Kim E-y, Kim M-s, Lee J-c, Jeong J, Kim B-S (2006) Leaching of copper from waste printed circuit boards using electrogenerated chlorine in hydrochloric acid solution. TMS 2006, 135th Annual Meeting & Exhibition, Recycling–General Sessions: Electronics Recycling, March 12–16, San Antonio, Texas, USA, pp. 929–933.

[21] Kim E-y, Kim M-s, Lee J-c, Jha MK, Yoo KK, Jeong J (2007) Effect of cuprous ions on copper leaching using electro-generated chlorine in hydrochloric acid solution. Bio-& hydrometallurgy 07, May 1–2, Falmouth, Cornwall, UK, pp. 1–16.

[22] Zhang Y, Liu S, Xie H, Zeng X, Li J (2012) Current status on leaching precious metals from waste printed circuit boards. *Procedia Environmental Sciences*, **16**: 560–568.

[23] Cui J, Zhang L (2008) Metallurgical recovery of metals from electronic waste: A review. *Journal of Hazardous Materials*, **158**(2–3): 228–256.

[24] Wan RY, Miller JD, Li J (2005) Thiohydrometallurgical processes for gold recovery. In: Innovations in Natural Resource Processing, In: Proceedings of the J.D. Miller Symposium, Ed. C.A. Young, J.J. Kellar, M.L. Free, J. Drelich, R.P. King, Society of Mining, Metallurgy, and Exploration, Inc. (SME) Colorado, USA (2006) pp. 223–244.

[25] https://en.wikipedia.org/wiki/Thiocyanate. Accessed 20 Sept. 2015.

[26] Tykodi RJ (1991) In praise of copper. *Journal of Chemical Education*, **68**: 106–109.

[27] Tudela D (1993) The reaction of copper (II) with thiocyanate ions. *Journal of Chemical Education*, **70**(2): 174. 10.1021/ed070p174.3.

[28] Aylmore MG, Muir DM (2001) Thiosulfate leaching of gold — a review. *Minerals Engineering*, **14**(2): 135–174.

[29] Chen J, Deng T, Zhu G, Zhao J (1996) Leaching and recovery of gold in thiosulfate based system — a research summary at ICM. *Transactions of the Indian Institute of Metals*, **49**(6): 841–849.

[30] Mishra D, Srivastava RR, Sahu KK, Singh TB, Jana RK (2011) Leaching of roast-reduced manganese nodules in NH_3–$(NH_4)_2CO_3$ medium. *Hydrometallurgy*, **109**(3–4): 215–220.

[31] Srivastava RR, Mishra AN, Singh VK (2013) Co-precipitation study of nickel and cobalt from ammoniacal leach liquor of laterite ore. *Journal of Metallurgy and Materials Science*, **55**(1): 67–72.

[32] Johnson BFG, Davis R (1973) Gold in Comprehensive Inorganic Chemistry (Eds. Bailar, J.C., Emeleus, H.J., Nyholm, R. Trotman-Dickenson, A.F.) Vol 5, Pergamon Press, p. 153.

[33] Pedraza AM, Villegas I, Freund PL, Chornik B (1988) Electro-oxidation of thiosulphate ion on gold: study by means of cyclic voltammetry and Auger electron spectroscopy. *Journal of Electroanalytical Chemistry and Interfacial Electrochemistry*, **250**(2): 443–449.

[34] Jiang T, Chen J, Xu S (1993) A kinetic study of gold leaching with thiosulfate. Hydrometallurgy: Fundamentals, Technology and Innovations. Soc Min, Metall Explor, Littleton, CO, USA, pp. 119–126.

[35] Meng X, Han KN (1993) The dissolution behaviour of gold in ammoniacal solutions in (Eds., Hiskey, J. B. and Warren, G.W.) Hydrometallurgy Fundamentals, Technology and Innovations SME, Littleton, Colorado, pp. 206–221.

[36] Tyurin NG, Kakowski IA (1960) Behaviour of gold and silver in oxidising zine of sulfide deposit Izu. Buz. Tsyv. *Metallurgy*, **2**: 6–13. (Russian).

[37] Ter-Arakelyan KA, Bagdasaryan KA, Oganyan AG, Mkrtchyan RT, Babayan GG (1984) On technological expediency of sodium thiosulphate usage for gold extraction from raw material.Izv. V.U.Z. Tsvetn. *Metallurgy*, 72–76.

[38] Tozawa K, Inui Y, Umetsu Y (1981) Dissolution of gold in ammoniacal thiosulfate solution. *AIME 110th Annual Meeting*, **A81–25**: 1–12.

[39] Byerley JA, Fouda SA, Rempel GL (1973) Kinetics and mechanism of the oxidation of thiosulfate ions by copper(II) ions in aqueous ammonia solution. *J Chem Soc*, Dalton Trans, 889–893. 10.1039/DT9730000889.

[40] Byerley JA, Fouda SA, Rempel GL (1973) The oxidation of thiosulfate in aqueous ammonia by copper (II) oxygen complexes. *Inorganic and Nuclear Chemistry Letters*, **9**: 879–883.

[41] Arima H, Fujita T, Yen WT (2004) Using Nickel as a Catalyst in Ammonium Thiosulfate Leaching for Gold Extraction. *Materials Transactions JIM*, **45**(2): 516–526.

[42] Ha VH, Lee Jc, Huynh TH, Jeong J, Pandey BD (2014) Optimizing the thiosulfate leaching of gold from printed circuit boards of discarded mobile phone. *Hydrometallurgy*, **149**: 118–126.

[43] Ha VH, Lee Jc, Jeong J, Hai HT, Jha MK (2010) Thiosulfate leaching of gold from waste mobile phones. *Journal of Hazardous Materials*, **178**: 1115–1119.

[44] Tripathi A, Kumar M, Sau DC, Agrawal A, Chakravarty S, Mankhand TR (2012) Leaching of gold from the waste mobile phone printed circuit boards (PCBs) with ammonium thiosulfate. *International Journal of Metallurgical Engineering*, **1**(2): 17–21.

[45] Camelino S, Rao J, Padilla RL, Lucci R (2015) Initial studies about gold leaching from printed circuit boards (PCBs) of waste cell phones. *Procedia Materials Science*, **9**: 105–112.

[46] Aylmore MG (2011) Treatment of a refractory gold-copper sulfide concentrate by copper ammoniacal thiosulfate leaching. *Minerals Engineering*, **14**(6): 615–637.

[47] Fleming CA (1998) The potential role of anion exchange resins in the gold industry. EPD Congress 1998. The Minerals, Metals and Materials Society, Warrendale, PA, USA, pp. 95–117.

[48] O'Malley GP (2001) The Elution of Gold from Anion Exchange Resins, International Patent WO 01/23626 A1 (April 2001).

[49] Grosse AC, Dicinoski GW, Shaw MJ, Haddad PR (2003) Leaching and recovery of gold using ammoniacal thiosulfate leach liquors (a review). *Hydrometallurgy*, **69**: 1–21.

[50] Langhans JW, Lei KPV, Carnahan TG (1992) Copper-catalysed thiosulfate leaching of low grade gold ores. *Hydrometallurgy*, **29**: 191–203.

[51] Jeffrey MI (2001) Kinetic aspects of gold and silver leaching in ammonia-thiosulfate solutions. *Hydrometallurgy*, **60**(1): 7–16.

[52] Flett DS, Derry R, Wilson JC (1983) Chemical study of thiosulfate leaching of silver sulfide, Transaction Institute of Mining and Metallurgy (Section C: Mineral Process. Extr. Metall), **92**: C216–223.

[53] Senanayake G (2004) Analysis of reaction kinetics, speciation and mechanism of gold leaching and thiosulfate oxidation by ammoniacal copper (II) solutions. *Hydrometallurgy*, **75**(1–4): 55–75.

[54] Petter PMH, Veit HM, Bernardes AM (2014) Evaluation of gold and silver leaching from printed circuit board of cellphones. *Waste Management*, **34**: 475–482.

[55] Petter PMH, Veit HM, Bernardes AM (2015) Leaching of gold and silver from printed circuit board of mobile phones. *Rem Revista Escola de Minas*, **68**(1): 61–68.

[56] Mine action (2000) Cyanide alternatives: Alternatives for Cyanide in the Gold Mining Industry. Great Basin Mine Watch, Reno.

[57] Yannopoulos, JC (1991) The Extractive Metallurgy of Gold. Van Nostrand Reinhol, USA.

[58] Munoz GA, Miller JD (2000) Noncyanide leaching of an auriferous pyrite ore from Ecuador. *Materials and Metallurgical Processing*, **17**(3): 198–204.

[59] Plaskin IN, Kozhukeva MA (1960) The solubility of gold and silver in thiourea. Dokl. Akad. Nauk SSSR 31, 671–674.

[60] Songina, OA, Ospanov KhK, Muldagalieva IKh, Sal'nikov SD (1971) Dissolution of gold with the use of thiourea in a hydrochloric acid medium. *Izvestiya Akademii Nauk Kazakhskoi SSR, Seriya Khimicheskaya*, **21**, 9–11.

[61] Li Jy, Xu Xl, Liu Wq (2012) Thiourea leaching gold and silver from the printed circuit boards of waste mobile phones. *Waste Management*, **32**(6): 1209–1212.

[62] Gurung M, Adhikari BB, Kawakita H, Ohto K, Inoue K, Alam S (2013) Recovery of gold and silver from spent mobile phones by means of acidothiourea leaching followed by adsorption using biosorbent prepared from persimmon tannin. *Hydrometallurgy*, **133**: 84–93.

[63] Huyhua JC, Zegarra CR, Gundiler IH (1989) A comparative study of oxidants on gold and silver dissolution in acidic thiourea solutions, in Precious Metals '89 (Ed. M.C. Jha, S.D. Hill, and El Guindy), pp. 287–303 (Minerals, Metals & Materials Society: Warrendale, 1986).

[64] Lacoste-Bouchet P, Deschenes G, Ghali E (1998) Thiourea leaching of a copper-gold ore using statistical design. *Hydrometallurgy*, **47**: 189–203.

[65] Groenewald T (1976) The dissolution of gold in acidic solutions of thiourea. *Hydrometallurgy*, **1**(3): 277–290.

[66] Habashi F (1969) Extractive Metallurgy Volume 1, General Principles. Science Publishers, Paris.

[67] Li J, Miller JD (1997) Thiourea decomposition by ferric sulfate oxidation in gold-leaching systems. Presentation at the SME Annual Meeting, February 24–27 SME, Denver, CO (1997) Preprint No. 97–146.

[68] Li J, Miller JD (2002) Reaction kinetics for gold dissolution in acid thiourea solution using formamidine disulfide as oxidant. *Hydrometallurgy*, **63**(3): 215–223.

[69] Birloaga I, Michelis ID, Ferella FBM, Veglio F (2013) Study on the influence of various factors in the hydrometallurgical processing of waste printed circuit boards for copper and gold recovery. *Waste Management*, **33**: 935–941.

[70] Behnamfard A, Salarirad MM, Veglio F (2013) Process development for the recovery of copper and precious metals from waste printed circuit boards with emphasize on palladium and gold leaching and precipitation. *Waste Management*, **33**: 2354–2363.

[71] Geoffroy N, Cardarelli F (2005) A method for leaching or dissolution gold from ores or precious metal scrap. *The Journal of The Minerals, Metals & Materials Society*, **57**(8): 47–50.

[72] Lee J-c, Bae MK, Srivastava RR, Kim S-k (2015) Separation of nitric acid and gold from aqua regia leach solution by TBP. In Proceeding of the 44th annual conference of The Korean Institute of Resource Recycling (Korean), pp. 10.

[73] Snoeyink PL, Jenkins D (1979) Water Chemistry. New York: John Wiley.

[74] Radulescu R, Filcenco-Olteanu A, Panturu E, Grigoras L (2008) New hydrometallurgical process for gold recovery. *Chem Bull "POLITEHNICA"* **53**: 135–139.

[75] Nicol MJ (1980) The anodic dissolution of gold. Part I. Oxidation in acidic solutions. *Gold Bulletin*, **13**: 46–55.

[76] Nicol MJ, Fleming CA, Paul RL (1987) The chemistry of the extraction of gold. In: The extractive metallurgy of gold in South Africa. G.G. Stanley (ed.), Vol. 2, The South African Institute of Mining and Metallurgy, Johannesburg.

[77] Finkelstein NP, Hancock RD (1974) A new approach to the chemistry of gold. *Gold Bulletin*, **7**(3): 72–77.

[78] Alex P, Mukherjee TK, Sundaresan M (1993) Leaching behavior of nickel in aqueous chlorine solutions and its application in the recovery of nickel from spent catalyst. *Hydrometallurgy*, **34**: 239–253.

[79] Lee MS, Nam SH (2009) Chemical equilibria of nickel chloride in HCl solution at 25°C. *Bulletin of the Korean Chemical Society*, **30**(10): 2203–2207.

[80] Kim Ey, Kim Ms, Lee Jc, Jha MK, Yoo K, Jeong J (2010) Leaching behavior of copper using electro-generated chlorine in hydrochloric acid solution. *Hydrometallurgy*, **100**: 95–102.

[81] Viñals J, Núñez C, Herreros O (1995) Kinetics of the aqueous chlorination of gold in suspended particles. *Hydrometallurgy*, **38**: 125–147.

[82] Chao MS (1968) Diffussion coefficients of hypochlorite, hypochlorous acid and chlorine in aqueous media by potentiometry. *Journal of The Electrochemical Society*, **115**(11): 1172–1174.

[83] Putnam GL (1944) Chlorine as a solvent in gold hydrometallurgy. *Engineering & Mining Journal*, **145**(3): 70–75.

[84] Isshiki Y, Satou H, Nagai H (2011) Recovery of precious metals in copper anode slime by chlorine leaching technology. *Journal of MMIJ*, **127**: 345–348.

[85] Brent HJ, Atluri VP (1998) Dissolution chemistry of gold and silver in different lixiviants. *Mineral Processing and Extractive Metallurgy Review*, **4**: 95–134.

[86] Baghalha M (2012) The leaching kinetics of an oxide gold ore with iodide/iodine solutions. *Hydrometallurgy*, **113–114**: 42–50.

[87] Marken F (2006) The electrochemistry of halogens. In: A. J. Bard. M. Stratmann, F. Scholz, C.J. Pickett (Eds.), Encyclopedia of Electrochemistry, Inorganic Chemistry, Chap 9, Vol. 7, Wiley-VCH (2006), pp. 291–297.

[88] Davis A, Tran T, Young DR (1993) Solution chemistry of iodide leaching of gold. *Hydrometallurgy*, **32**: 143–159.

[89] Xu Q, Chen DH, Chen L, Huang MH (2009) Iodine leaching process for recovery of gold from waste PCB. *Chinese Journal of Environmental Engineering*, **3**(5): 911–914.

[90] Xu Q, Chen DH, Chen L, Huang MH (2010) Gold leaching from waste printed circuit board by iodine process. *Nonferrous Metals* **62**(3): 88–90.

[91] Yin J, Zhan S, Xu H (2014) Comparison of leaching processes of gold and copper from printed circuit boards of waste mobile phone. *Advanced Materials Research*, **955–959**: 2743–2746.

[92] Xiu FR, Qi Y, Zhang FS (2015) Leaching of Au, Ag, and Pd from waste printed circuit boards of mobile phone by iodide lixiviant after supercritical water pre-treatment. *Waste Management*, **41**: 134–141.

[93] David A, Tran T (1991) Gold dissolution in iodide electrolytes. *Hydrometallurgy*, **26**: 163–177.

Chapter 3
Electroless Displacement Deposition of Gold from Aqueous Source — Recovery from Waste Electrical and Electronic Equipment (WEEE) using Waste Silicon Powder

Kenji Fukuda and Shinji Yae

3.1 Introduction

Electroless displacement deposition, that is, cementation in hydrometallurgy, is one of the common methods to recover any dissolved metals from aqueous solutions [1–47]. More precisely, cementation is described as the electrochemical precipitation of metals from aqueous solution including those ions by using another more electropositive (less noble) metal as a sacrificial anode. One of the common and traditional cementation processes is refining gold ore into pure metal (e.g. Merrill–Crowe process) [1–7]. The cementation has several advantages including the use of inexpensive base metals such as zinc and iron, simplicity of process adding base metal into an aqueous solution, selectivity of recovering metals by selecting base metal

K. Fukuda and S. Yae*
Department of Chemical Engineering and Materials Science
Graduate School of Engineering, University of Hyogo
2167 Shosha, Himeji, Hyogo 671–2280, Japan
e-mail: *yae@eng.u-hyogo.ac.jp

having electrochemical potential between noble metals and impurities, and effectiveness of noble metal recovery even from leaching solutions. Thus, this method has been successfully used not only in the refining process of natural resources but also in the recovering process of noble metals from artificial waste water produced by such industries as metal plating, surface finishing, and electronics. Recently, it is also used in the process of noble metal recovering from waste electrical and electronic equipment (WEEE) [48], urban mines in other word.

The electroless displacement deposition of noble metals on silicon (Si) substrates has been attracting much attention from both the viewpoints of semiconductor electrochemistry and electronics industry [49–56]. It produces noble metal nanoparticles sparsely scattered on silicon surfaces at the widely controlled particle density between $10^6 \, cm^{-2}$ and $10^{11} \, cm^{-2}$ [55–58]. It is a simple process involving only the immersion of substrates into metal–salt solutions including fluoride species such as hydrofluoric acid or ammonium fluoride. This deposition method has been studied as deposition mechanism of metal impurity affecting silicon semiconductor device performance [50, 51] and for various applications, such as surface-enhanced infrared absorption reflection spectroscopy [59–61], defect detection of silicon wafers [62–64], and deposition of noble metal catalysts on silicon. The catalysts are widely used for metal assisted etching of silicon to produce porous layers and nanowires [65–72], solar hydrogen evolution using photoelectrochemical solar cells [73–75], and autocatalytic electroless deposition of metal thin films on silicon [76–79]. Recently, we reported that this electroless displacement deposition can efficiently recover noble metals from aqueous solutions by using waste silicon powder, which is produced as sawdust by slicing silicon ingots to wafers and dicing silicon wafers to integrated circuit (IC) chips [80–82]. This recovering method is not only efficient (high reaction and recovery rates), but also unable to recover less noble metals than copper, thus able to select noble metals and copper from the mixture solution of various metal ions.

In this chapter, we review the principle of electroless displacement deposition and conventional cementation processes of gold, and describe our novel method to recover gold from the WEEE using waste silicon powder.

3.2 Electroless Displacement Deposition [47,83]

The overall reaction of electroless displacement deposition is generally given by following equation:

$$zM_l + mM_r{}^{z+} \rightarrow zM_l{}^{m+} + mM_r \tag{1}$$

and cell formula,

$$M_l \big| M_l{}^{m+}(aq) \big| \big| M_r{}^{z+}(aq) \big| M_r, \tag{2}$$

where M_l is the less noble metal than metal M_r dissolved as ion $M_r{}^{z+}$ in solution. This is a local galvanic cell, as schematically shown in Fig. 1, consisting of following reactions:

$$\text{Local anode reaction (Oxidation): } M_l \rightarrow M_l{}^{m+} + me^- \tag{3}$$

and

$$\text{Local cathode reaction (Reduction): } M_r{}^{z+} + ze^- \rightarrow M_r. \tag{4}$$

Metal M_l is dissolving into the solution [Eq. (3)], thus supplying electrons necessary for the reduction of $M_r{}^{z+}$ ions, that is, M_r metal deposition [Eq. (4)]. The potentials and electron flow of the local galvanic cell are

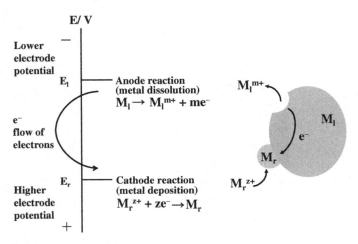

Fig. 1   Schematics of local galvanic cell reactions in electroless displacement deposition

The Recovery of Gold from Secondary Sources

schematically shown in Fig. 1, where E_r is the potential of the local cathode reaction of $M_r{}^{z+}/M_r$ and E_l is the potential of the local anode reaction of $M_l{}^{m+}/M_l$. At equilibrium, the electrode potential (E) of such half-cell reaction as Eq. (4), $M_r{}^{z+} + ze^- = M_r$, is expressed as a function of the metal ion activity in the solution in accordance with the following Nernst equation:

$$E = E^0 + \frac{RT}{zF} \ln a_{M^{z+}}, \tag{5}$$

where E^0 is the standard electrode potential of the reaction, a is the activity, R is the gas constant (8.3145 J K^{-1} mol^{-1}), T is the temperature, and F is the Faraday constant (96485 C mol^{-1}). The standard potential (E^0) means equilibrium potential under the conditions that all activities of the

Table 1 Standard potential (E^0) of M^{z+}/M and M^{z+}/M^{m+} in aqueous solution [84]

Reaction	E^0/V vs. SHE
$Al^{3+} + 3e^- = Al$	−1.676
$Zn^{2+} + 2e^- = Zn$	−0.7626
$Fe^{2+} + 2e^- = Fe$	−0.44
$Co^{2+} + 2e^- = Co$	−0.277
$Ni^{2+} + 2e^- = Ni$	−0.228
$Sn^{2+} + 2e^- = Sn$	−0.138
$2H^+ + 2e^- = H_2$	**0.0000**
$Cu^{2+} + 2e^- = Cu$	+0.340
$Cu^+ + e^- = Cu$	+0.520
$Rh^{3+} + 3e^- = Rh$	+0.76
$Ag^+ + e^- = Ag$	+0.7991
$Pd^{2+} + 2e^- = Pd$	+0.915
$Ir^{3+} + 3e^- = Ir$	+1.156
$Pt^{2+} + 2e^- = Pt$	+1.188
$Au^{3+} + 3e^- = Au$	+1.52
$Au^+ + e^- = Au$	+1.83
$Cu^{2+} + e^- = Cu^+$	+0.159
$Fe^{3+} + e^- = Fe^{2+}$	+0.771
$Ag^{2+} + e^- = Ag^+$	+1.980
$Au^{3+} + 2e^- = Au^+$	+1.36

substances involved in the electron transfer reaction are unity. Table 1 shows the standard potential of several reactions related to displacement metal deposition for cementation of noble metals in aqueous solution.

The thermodynamic criterion for spontaneity (feasibility) of a chemical or electrochemical reaction is that the reaction Gibbs energy, $\Delta_r G$, has a negative value. The reaction Gibbs energy of an electrochemical cell reaction can be calculated from electromotive force, ΔE, that is the electrostatic potential of electrode written on your right-hand side in a cell formula (such as M_r in Eq. (2)) with respect to that of the left (such as M_l in Eq. (2)):

$$\Delta_r G = -pF\Delta E, \tag{6}$$

where p is the number of electrons transferred by the reaction (such as mz (the product of m and z) for the reaction of Eq. (1)). If the activities of all species are unity, $\Delta_r G$ and ΔE are the standard reaction Gibbs energy, $\Delta_r G^0$, and standard electromotive force, ΔE^0, as follows:

$$\Delta_r G^0 = -pF\Delta E^0. \tag{7}$$

A spontaneous reaction must have a positive electromotive force. The electromotive force can be calculated from the electrode potential E for the half-cell reactions using the expression in Eq. (2):

$$\Delta E = E_r - E_l. \tag{8}$$

When E_r is more positive than E_l, the positive electromotive force is obtained, thus the reaction is spontaneous (feasible) in a direction from left to right of Eq. (1). This is the displacement deposition of metal M_r using less noble metal M_l.

For example, ΔE^0 for the electroless displacement deposition of gold on a zinc substrate is calculated. The reaction is expressed by the following equations:

$$\text{Local cathode reaction: } Au^{3+} + 3e^- \rightarrow Au, \tag{9}$$

$$\text{Local anode reaction: } Zn \rightarrow Zn^{2+} + 2e^-, \tag{10}$$

$$\text{Overall reaction: } 3Zn + 2Au^{3+} \rightarrow 3Zn^{2+} + 2Au. \tag{11}$$

We have

$$E^0 \left(\frac{Au^{3+}}{Au} \right) = +1.52 \text{ V vs. SHE,} \tag{12}$$

$$E^0 \left(\frac{Zn^{2+}}{Zn} \right) = -0.763 \text{ V vs. SHE} \tag{13}$$

and obtain

$$\Delta E^0 = E^0 \left(\frac{Au^{3+}}{Au} \right) - E^0 \left(\frac{Zn^{2+}}{Zn} \right) = +1.52 - (-0.763) = 2.283 \text{ V,} \tag{14}$$

$$\Delta_r G^0 \doteqdot -1322 \text{ kJ mol}^{-1}. \tag{15}$$

The same value of ΔE^0 is obtained experimentally. Since ΔE^0 is positive and $\Delta_r G^0$ is negative, the overall electroless displacement deposition reaction proceeds spontaneously from left to right under the standard conditions.

At the equilibrium of overall reaction Eq. (11), $3Zn + 2Au^{3+} = 3Zn^{2+} + 2Au$, the electrode potentials of Zn^{2+}/Zn and Au^{3+}/Au systems are equal ($\Delta E = 0$) and the equilibrium constant, $K = a_{Zn^{2+}}{}^3/a_{Au^{3+}}{}^2$, is calculated as follows:

$$E \left(\frac{Au^{3+}}{Au} \right) = E \left(\frac{Zn^{2+}}{Zn} \right), \tag{16}$$

$$E^0 \left(\frac{Au^{3+}}{Au} \right) + \frac{RT}{3F} \ln a_{Au^{3+}} = E^0 \left(\frac{Zn^{2+}}{Zn} \right) + \frac{RT}{2F} \ln a_{Zn^{2+}}, \tag{17}$$

$$\ln K = -\frac{\Delta_r G^0}{RT} = -\frac{-1322 \times 10^3}{8.3145 \times 298} = 534. \tag{18}$$

This large value of equilibrium constant means that the concentration of gold ions in the solution is extremely low, thus gold is effectively recovered by the cementation using zinc. Table 2 shows the standard electromotive force and the reaction Gibbs energy of several displacement reactions. In case ΔE^0 is less than 0.3 V, there is a possibility that the reverse reaction occurs due to the difference of real equilibrium potential (formal potential)

Table 2 Standard electromotive force (ΔE^0) and the reaction Gibbs energy ($\Delta_r G^0$) of overall reaction [84]

Overall reaction	$\Delta E^0/V$	$-\Delta_r G^0/kJ\,mol^{-1}$
$2Au^{3+} + 3Cu = 2Au + 3Cu^{2+}$	1.183	684
$2Au^{3+} + 3Fe = 2Au + 3Fe^{2+}$	1.960	1135
$2Au^{3+} + 3Zn = 2Au + 3Zn^{2+}$	2.283	1322
$Au^{3+} + Al = Au + Al^{3+}$	3.196	925
$2Au^{+} + Cu = 2Au + Cu^{2+}$	1.493	288
$2Au^{+} + Zn = 2Au + Zn^{2+}$	2.593	500
$3Au^{+} + Al = 3Au + Al^{3+}$	3.506	1015
$2Ag^{+} + Fe = 2Ag + Fe^{2+}$	1.239	239
$2Ag^{+} + Zn = 2Ag + Zn^{2+}$	1.562	301
$Pt^{2+} + Fe = Pt + Fe^{2+}$	1.628	314
$Pt^{2+} + Zn = Pt + Zn^{2+}$	1.951	376
$Pd^{2+} + Fe = Pd + Fe^{2+}$	1.355	261
$Pd^{2+} + Zn = Pd + Zn^{2+}$	1.678	324
$Cu^{2+} + Fe = Cu + Fe^{2+}$	0.777	150
$Cu^{2+} + Zn = Cu + Zn^{2+}$	1.100	212
$Ni^{2+} + Zn = Ni + Zn^{2+}$	0.535	103

from the standard potential. When there is a difference of more than 0.3 V, it does not care to use the standard potential for considering spontaneity (feasibility) of displacement deposition.

In the solution containing the ligand, the value of standard potential varies greatly. Table 3 shows standard potential for metal complexes. When the ligand L^{y-} and metal ion M^{z+} forms complex ion $ML_n^{(z-ny)}$, the reduction reaction and potential are expressed as follows:

$$ML_n^{(z-ny)} + ze^- = M + nL^{y-}, \tag{19}$$

$$E = E_L^0 + \frac{RT}{zF} \ln \frac{a_{ML_n^{(z-ny)}}}{a_{L^-}^n}, \tag{20}$$

where E_L^0 is standard electrode potential of Eq. (19). The standard electrode potential of M^{z+}/M system ($M^{z+} + ze^- \rightarrow M$), E^0, and E_L^0 are related by

Table 3 Standard potential for metal complexes [84]

Reaction	E^0/V vs. SHE
$Au(CN)_2^- + e^- = Au + 2CN^-$	-0.595
$Au(S_2O_3)_2^{3-} + e^- = Au + 2S_2O_3^{2-}$	$+0.153$
$Au[CS(NH_2)_2]_2^+ + e^- = Au + 2CS(NH_2)_2$	$+0.380$
$Au(NH_3)_2^+ + e^- = Au + 2NH_3$	$+0.563$
$AuI_2^- + e^- = Au + 2I^-$	$+0.578$
$Au(SCN)_2^- + e^- = Au + 2SCN^-$	$+0.662$
$AuBr_2^- + e^- = Au + 2Br^-$	$+0.960$
$AuCl_2^- + e^- = Au + 2Cl^-$	$+1.154$
$Au(NH_3)_4^{3+} + 3e^- = Au + 4NH_3$	$+0.325$
$AuI_4^- + 3e^- = Au + 4I^-$	$+0.56$
$Au(SCN)_4^- + 3e^- = Au + 4SCN^-$	$+0.636$
$AuBr_4^- + 3e^- = Au + 4Br^-$	$+0.854$
$AuCl_4^- + 3e^- = Au + 4Cl^-$	$+1.002$
$AuI_4^- + 2e^- = AuI_2^- + 2I^-$	$+0.55$
$Au(SCN)_4^- + 2e^- = Au(SCN)_2^- + 2SCN^-$	$+0.623$
$AuBr_4^- + 2e^- = AuBr_2^- + 2Br^-$	$+0.802$
$AuCl_4^- + 2e^- = AuCl_2^- + 2Cl^-$	$+0.926$
$IrCl_6^{2-} + 4e^- = Ir + 6Cl^-$	$+0.86$
$IrCl_6^{3-} + 3e^- = Ir + 6Cl^-$	$+0.86$
$PdCl_4^{2-} + 2e^- = Pd + 4Cl^-$	$+0.60$
$PtCl_6^{2-} + 4e^- = Pt + 6Cl^-$	$+0.744$
$PtCl_4^{2-} + 2e^- = Pt + 4Cl^-$	$+0.758$
$Cu(CN)_2^- + e^- = Cu + 2CN^-$	-0.44
$CuCl + e^- = Cu + Cl^-$	$+0.121$
$Zn(CN)_4^{2-} + 2e^- = Zn + 4CN^-$	-1.34
$Zn(NH_3)_4^{2-} + 2e^- = Zn + 4NH_3$	-1.04
$[Fe(CN)_6]^{4-} + 2e^- = Fe + 6CN^-$	-1.16
$[Fe(CN)_6]^{3-} + e^- = [Fe(CN)_6]^{4-}$	$+0.3610$

the overall stability constant β_n:

$$M^{z+} + nL^{y-} = ML_n^{(z-ny)}, \tag{21}$$

$$\beta_n = \frac{a_{ML_n^{(z-ny)}}}{a_{M^{z+}} a_{L^{y-}}^n}. \tag{22}$$

When hydrated ions are in equilibrium with the complex ions, the following relationship holds:

$$E^0 + \frac{RT}{zF} \ln a_{M^{z+}} = E_L^0 + \frac{RT}{zF} \ln \frac{a_{ML_n^{(z-ny)}}}{a_{L^{y-}}^n}, \tag{23}$$

$$E_L^0 = E^0 - \frac{RT}{zF} \ln \beta_n. \tag{24}$$

The standard electrode potential of metal is reduced by addition of a complexing agent in the solution, and it depends on the stability constant of complex.

The following equation expresses the electroless displacement deposition reaction of gold by zinc from a cyanide solution:

$$2[Au(CN)_2]^- + Zn \rightarrow [Zn(CN)_4]^{2+} + 2Au. \tag{25}$$

The half-cell reactions of local galvanic cell are as follows:

Local cathode reaction:

$$Au(CN)_2^- + e^- \rightarrow Au + 2CN^- \quad E^0 = -0.595\,V \text{ vs. SHE}, \tag{26}$$

Local anode reaction:

$$Zn + 4CN^- \rightarrow Zn(CN)_4^{2-} + 2e^- \quad E^0 = -1.34\,V \text{ vs. SHE}. \tag{27}$$

In comparison with the cases that included no cyanide ion, the standard potentials of local reactions are lower (less noble). The potential shift of gold by addition of cyanide ions is much larger than that of zinc. Thus, the standard electromotive force of this electroless displacement deposition reaction in a cyanide solution is 0.745 V, much lower than 2.21 V in a non-cyanide solution.

A large potential difference between the local electrode reactions of electroless displacement deposition induces precipitation of impurities, whose electrode potentials are in the potential difference. For example, reduction of gold ions using zinc is likely to precipitate other metals such as silver, copper, nickel, and so on. In addition, such reactions may occur on the cathode as follows:

$$O_2 + 4H^+ + 4e^- \rightarrow 2H_2O \ E^0 = +1.229 \text{ vs. SHE}, \qquad (28)$$

$$2H^+ + 2e^- \rightarrow H_2 \ E^0 = 0.000 \text{ V vs. SHE}, \qquad (29)$$

$$2H_2O + 2e^- \rightarrow 2OH^- + H_2 \ E^0 = -0.828 \text{ vs. SHE}. \qquad (30)$$

Since oxygen/water system (Eq. (28)) has a large positive potential, oxygen is reduced in preference to metal ions. Hydrogen evolution expressed by Eqs. (29) and (30) also affects displacement metal deposition. Both the oxygen reduction and hydrogen evolution depends on the potential and catalytic activity (overvoltage) of metal, as well as dissolved oxygen concentration and pH of solution. In the cases of such noble metal having low catalytic activity of hydrogen evolution as gold and silver, the metal deposition precedes the hydrogen evolution. The increase of solution pH is effective to depress the hydrogen evolution, but there is a risk of hydrolysis of the metal ions.

3.3 Conventional Cementation of Gold Using Metals

3.3.1 *Zinc* [1–17]

Cementation, that is, displacement deposition using zinc for gold recovery is one of the best known reducing processes in mineral processing. The cementation was introduced commercially for the treatment of cyanide leach solution of gold in the 1890s, and subsequently applied widely in industries as an alternative to electrowinning. A Merrill–Crowe process, in which the main reaction is expressed by Eq. (25), evolved as the efficient recovering method of gold from cyanide solutions and has been applied worldwide [1–6].

Leaching of gold using cyanide is

$$4Au + 8CN^- + O_2 + 2H_2O \rightarrow 4Au(CN)_2^- + 4OH^-. \qquad (31)$$

This reaction consists of following local cell reactions:

$$\text{Local anode reaction: } Au + 2CN^- \rightarrow Au(CN)_2^- + e^-, \qquad (32)$$

$$\text{Local cathode reaction: } O_2 + 2H_2O + 4e^- \rightarrow 4OH^-. \qquad (33)$$

Although the local cathode reaction of cementation of gold using zinc from a cyanide solution is expressed by Eq. (26), several side reactions can proceed as mentioned in the preceding section in solution. The typical side reaction is the reduction of dissolved oxygen expressed in Eqs. (28) and (34) in an aqueous solution of high pH.

$$O_2 + H_2O + 2e^- \rightarrow OH^- + HO_2^-. \tag{34}$$

Thus, the removal of dissolved oxygen is generally carried out by vacuum deaeration in Merrill–Crowe process.

The local anode, whose main reaction is expressed by Eq. (27), also has such side reactions in the high pH solution as follows [7]:

$$Zn + 2OH^- \rightarrow Zn(OH)_2 + 2e^-, \tag{35}$$

$$Zn + 2OH^- \rightarrow HZnO_2^- + H^+ + 2e^-, \tag{36}$$

$$Zn + 4OH^- \rightarrow ZnO_2^{2-} + 2H_2O + 2e^-. \tag{37}$$

The formation of zinc hydroxide is undesirable because it covers and passivates the surface of zinc, and thus impedes the gold deposition [2]. It was reported that addition of lead to the solution is effective to inhibit passivation of zinc surface [2–5,9].

For the efficient recovery, it is important to consider not only the dissolved oxygen [5, 7], but also several parameters, such as cyanide concentration [2, 5–7], pH [5, 7], zinc concentration [6], zinc particle size [2, 7], and solution temperature [2, 7, 9].

The concentration of cyanide must be optimized along with the addition of zinc powder. At low cyanide concentration, the gold deposition is terminated apparently due to zinc hydroxide formation (Eq. (35)) [2]. Although high cyanide concentration reduces the zinc hydroxide formation, it increases zinc consumption through zinc dissolution with hydrogen evolution (Eq. (30)), which is one of the cathode side reactions.

A decrease in particle size that is an increase in surface area of zinc powder improves the reaction rate of displacement deposition (cementation) due to surface reaction. However, it accelerates the side reaction of zinc dissolution. The solution temperature effects in a similar manner. High temperature leads high reaction rates of both main and side reactions.

Cyanide is a popular and effective reagent (ligand) for leaching gold, but it is toxic. Therefore, ammoniacal thiosulphate leaching was proposed as a leaching using low-toxic reagent. Gold cementation reaction using zinc from ammoniacal thiosulphate is as follows:

$$2Au(S_2O_3)_2^{3-} + Zn \rightarrow 2Au + Zn^{2+}4S_2O_3^{2-}. \tag{38}$$

3.3.2 *Aluminum* [4, 18]

The use of aluminum for cementation of gold using aluminum from alkaline cyanide solutions was originally proposed in 1890s [4]. Despite some advantages over zinc, the process using aluminum has not been applied widely due to high cost. The reaction of electroless displacement deposition of gold using aluminum from a cyanide solution is as follows:

Overall reaction

$$3Au(CN)_2^- + Al + 4OH^- = 3Au + 6CN^- + AlO_2^- + 2H_2O. \tag{39}$$

Several other reactions may proceed as side reactions, but they can be controlled by specific solution conditions. This cementation reaction yields two free cyanide ions for each gold yielding. In contrast, the same cementation reaction using zinc yields no free cyanide ion, and the side reaction of zinc dissolution actually consumes additional four free cyanide ions as Eq. (27). The regeneration of cyanide is an important advantage of the cementation using aluminum.

3.3.3 *Copper* [19–23]

The cementation of gold using copper have been reported previously. Leaching of gold in ammoniacal thiosulphate solutions often uses cupric (Cu^{2+}) ions as a catalytic oxidizing agent of the following reactions:

$$Au + 5S_2O_3^{2-} + Cu(NH_3)_4^{2+}$$
$$\rightarrow Au(S_2O_3)_2^{3-} + 4NH_3 + Cu(S_2O_3)_3^{5-} \tag{40}$$

$$2Cu(S_2O_3)_3^{5-} + 8NH_3 + \frac{1}{2}O_2 + H_2O$$
$$= 2Cu(NH_3)_4^{2+} + 2OH^- + 6S_2O_3^{2-}. \tag{41}$$

The cupric ions oxidize gold, and then produced cuprous ions (Cu^+) are oxidized by dissolved oxygen or other oxidizing agents. The cementation of gold from the ammoniacal thiosulphate solution (Eq. (42)) efficiently proceeds by using copper metal as a reducing agent [23]. The local cathode and anode reactions of cementation are enhanced by the presence of copper species, metallic or cuprous, and ammonia in the solution, respectively.

$$Au(S_2O_3)_2^{3-} + S_2O_3^{2-} + Cu \rightarrow Cu(S_2O_3)_3^{5-} + Au. \qquad (42)$$

3.4 Recovery of Noble Metals using Waste Silicon Powder

3.4.1 *Introduction*

Semiconductor silicon, which is ultrapure over 99.9999999% and single or multi-crystalline, is one of the key materials for present human beings. It is manufactured in several 100,000 tons per year for semiconductor and photovoltaic industries. These industries produce waste silicon powder as sawdust during processes of slicing silicon ingots to wafers and dicing silicon wafers to IC chips. About 60% of silicon ingots is disposed as waste silicon powder in sludge. The standard electrode potentials of such silicon related reactions are highly negative as expressed in the following [72]:

$$SiF_6^{2-} + 2HF + H_2 + 2e^- = Si + 4HF_2^- \quad E^0 = -1.2 \text{ vs. SHE} \quad (43)$$

and/or

$$SiO_2 + 4H^+ + 4e^- = Si + 2H_2O \quad E^0 = -0.84 \text{ vs. SHE}. \qquad (44)$$

The waste silicon powder is expected to be a reducing agent as base metals for the cementation of not only gold but also many noble metals.

3.4.2 *Electroless displacement deposition of noble metal nanoparticles on silicon*

We previously reported that noble metal nanoparticles are deposited on silicon substrates by the electroless displacement deposition [55–58]. This deposition is a simple process involving immersion of silicon substrates into metal–salt solutions including hydrofluoric acid as schematically indicated in Fig. 2. It can produce such noble metal nanoparticles as gold (Au),

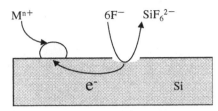

Fig. 2 Schematic illustration of displacement deposition of metal on silicon in fluoride containing solution

silver (Ag), palladium (Pd), rhodium (Rh), and platinum (Pt). Figure 3 shows scanning electron microscopic (SEM) images of nanoparticle-deposited silicon surfaces [55, 56, 73–75]. The nanoparticles are sparsely scattered on silicon surfaces at the wide particle density. Figure 4 indicates that the particle density is well controlled between 10^6 cm^{-2} and 10^{11} cm^{-2} [55, 56]. This deposition is effective not only for single-crystalline silicon wafers but also for multi-crystalline wafers as well as micro-crystalline and amorphous silicon thin films [73–75].

This deposition is a local galvanic reaction consisting of local cathodic deposition of metals and local anodic dissolution of silicon. Figure 5 shows the band diagram and standard electrode potential of noble metals. Efficient recovery of noble metals using the galvanic reaction is expected owing to the large electrochemical potential difference between noble metal deposition and silicon dissolution.

3.4.3 *Experimental*

Standard experimental conditions are as follows. Silicon powder was put in a mixture solution of 1 mmol dm^{-3} metal salts and 0.2 mol dm^{-3} hydrofluoric acid. The amount of added silicon powder was corresponding to 0.20 mol per 1.00 dm^3 of the solution. We will hereafter explain this amount as 0.20 mol dm^{-3} of silicon powder. We used three kinds of silicon powder. One was commenced silicon powder, whose particle size was smaller than 0.045 mm (mean diameter: 0.016 mm, purity: 99.99%, Kojundo Chemical Laboratory). We used pure silicon powder because the silicon powder of such industrial wastes as sawdust of slicing crystalline ingots and dicing was expected to have high purity. The other two kinds of silicon powder we used were sawdust silicon, which was produced by industrially slicing

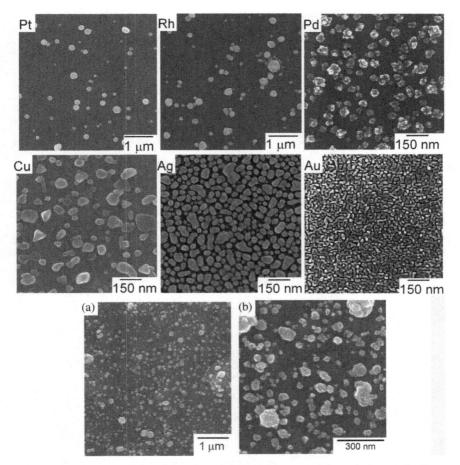

Fig. 3 SEM images of nanoparticle-deposited silicon surfaces. platinum, rhodium, palladium, copper, silver, and gold: single-crystalline n-type silicon wafers after immersion in each metal salt solution including hydrofluoric acid [55, 56]; and (a) and (b): multi-crystalline n-type silicon wafer [73] and microcrystalline silicon thin film electrode [74] after immersion in a hexachloroplatinic (IV) acid solution including hydrofluoric acid, respectively

single-crystalline ingots with multi-wire saws. The sawdust silicon was used after washing with water. The examined solution contained one or more of the following metal salts: hydrogen tetrachloroaurate(III) tetrahydrate ($HAuCl_4 \cdot 4H_2O$), silver(I) nitrate ($AgNO_3$), palladium(II) chloride ($PdCl_2$),

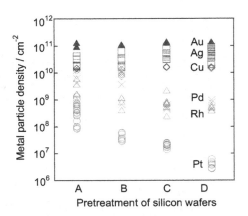

Fig. 4 Particle density of various metal nanoparticles deposited by electroless displacement deposition on single-crystalline n-type silicon wafers. Pretreatment of silicon wafers: A using CP4A solution, B RCA method, and C and D immersion in nitric acid after A and B, respectively [55, 56]

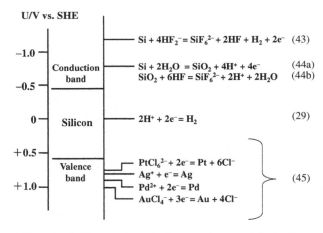

Fig. 5 Band diagram of silicon and standard electrode potential of noble metals [84, 85]

rhodium(III)chloride ($RhCl_3$), hydrogen hexachloroplatinate(IV) hexahydrate ($H_2PtCl_6 \cdot 6H_2O$), osmium(III) chloride ($OsCl_3$), ruthenium(III) chloride ($RuCl_3$), potassium hexachloroiridate(IV) (K_2IrCl_6), copper(II) chloride dihydrate ($CuCl_2 \cdot 2H_2O$), nickel(II) chloride hexahydrate ($NiCl_2 \cdot 6H_2O$), iron(III) chloride hexahydrate ($FeCl_3 \cdot 6H_2O$), and cobalt(II) chloride hexahydrate ($CoCl_2 \cdot 6H_2O$). After stirring the mixture solution of metal salts, hydrofluoric acid, and silicon powder at 200 rpm and a solution temperature of 298 K, the solution was filtered. The concentration of the metal in the solution was measured by an inductively coupled plasma-atomic emission spectrometry (ICP-AES, Hitachi High-technologies, SPS 7800). In this study, the recovery of noble metals was determined by the removal rate of the noble metal ions in the solution. The recovery (removal rate) was calculated from the change of the metal concentration of the examined solution before and after the reaction using the following equation:

$$\text{Recovery } (\%) = \frac{\text{initial concentration} - \text{concentration of filtrate}}{\text{initial concentration}} \times 100.$$

(46)

The deposits on the silicon were inspected with an scanning electron microscope (SEM, JEOL, JSM-7001FA) with an energy dispersive X-ray spectrometer (EDS). Sawdust silicon was identified with X-ray diffraction (XRD, Rigaku, Ultima IV).

3.4.4 *Recovery of noble metals using silicon powder*

Figure 6 shows the process of gold recovery. Silicon powder of 2.00 mol dm^{-3} was put in a mixture solution of 25 mmol dm^{-3} $HAuCl_4$ and 1.00 mol dm^{-3} hydrofluoric acid. The mixture solution was stirred for 10 min. A colorless filtrate solution, unlike the initial yellowish

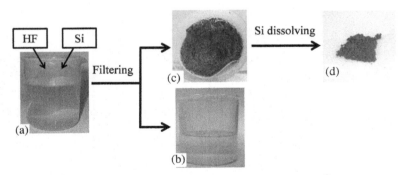

Fig. 6 Recovery of gold from aqueous solution. (a) 25 mmol dm^{-3} $HAuCl_4$ solution, (b) and (c) filtrate and residue after recovery, and (d) residue after silicon dissolving with hydrofluoric acid and nitric acid solution

solution (Fig. 6(a)), and residue were obtained by filtration of the solution (Figs. 6(b) and 6(c)). The gold concentration in the filtrate was 0.4 mol dm^{-3}. This concentration change corresponds to 99.999% recovery from 4840 mg dm^{-3} to 0.07 mg dm^{-3} of gold in the solution. The residue (Fig. 6(c)) was added in a mixture solution of 4.2 mol dm^{-3} hydrofluoric acid and 9.4 mol dm^{-3} nitric acid (HNO$_3$) to dissolve the silicon powder. After the dissolution for 30 min, most of the residues disappeared, and a small amount of brown powder was obtained by filtering the solution (Fig. 6(d)). The EDS spectrum of the powder has a large peak assigned to the Au-Mα line at 2.12 keV and no peak assigned to Si-Kα at 1.74 keV (Fig. 7). These results indicate that the present method can efficiently recover pure gold powder from a solution containing gold ions.

Figure 8 shows the recovery plots of noble metals vs. treatment time, which is stirring time of the mixture solution of 1 mmol dm^{-3} metal salt, 0.15 mol dm^{-3} hydrofluoric acid, and 0.2 mol dm^{-3} silicon powder. Each metal was recovered from a solution containing a single chloride or chlorido-complex metal salt except silver nitride. The recovery rate descends in the order of silver, gold, palladium, rhodium, platinum, osmium, and ruthenium. In particular, silver, gold, and palladium were efficiently recovered at 99.9% in 1 min. However, iridium could not be recovered by the treatment for 600 s. Tending the treatment time and modification of silicon powder with metal nuclei were examined for recovering of iridium. Figure 9 shows the recovery of iridium. The bare

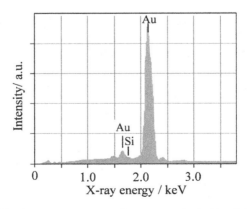

Fig. 7   EDX spectrum of brown powder in Fig. 6(d)

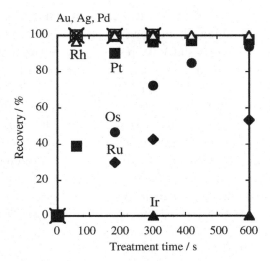

Fig. 8 Recovery of noble metals from aqueous solution of 1 mmol dm^{-3} single metal salt, 0.15 mol dm^{-3} hydrofluoric acid, and 0.2 mol dm^{-3} silicon powder. gold: white circles, silver: white squares, palladium: crosses, rhodium: white triangles, platinum: black squares, osmium: black circles, ruthenium: black diamonds, and iridium: black triangles

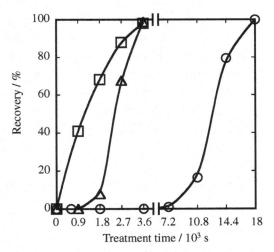

Fig. 9 Recovery of iridium using gold-deposited silicon powder (triangles), iridium-deposited silicon powder (squares), and bare silicon powder (circles)

silicon powder recovers 99.94% of iridium by treatment for 18,000 s. The iridium and gold modified silicon powders prepared by the same treatment as the recovery for 18,000 s and 360 s, respectively, recovered iridium by the treatment for 3,600 s, much shorter than that for the bare silicon powder. The nucleation activity of the noble metal displacement deposition on silicon wafers depends on the kind of metal. The activity of platinum is much lower than that of silver, and silver nuclei activate platinum deposition. In this case, platinum deposited not on silicon directly but on silver nuclei. These results indicate that the nucleation activity of iridium is much lower than that of other noble metals, but iridium's highly positive deposition potential promotes its displacement deposition onto catalytic nuclei on silicon. Figure 10 shows the SEM images of silicon powder surface after immersion in each metal ion solution. Fine metal particles were unevenly deposited on the silicon powder surfaces.

To recover noble metals more efficiently, the dependence of recovery on quantity of silicon, shape of silicon, hydrofluoric acid concentration, and solution temperature were investigated. Figure 11 shows the dependence of platinum recovery on quantity (a) and size (b) of silicon powder. The rate of recovery increased as the quantity of silicon powder increased and the size of silicon particles decreased. These indicate that the rate of recovery increases with surface area of silicon powder added in the solution, and the electroless displacement deposition is surface reaction (Fig. 5). Figure 12 shows the dependences of the platinum and ruthenium recovery on the hydrofluoric acid concentration (a) and temperature (b) of the solution, respectively. The recovery rates increased with hydrofluoric acid concentration and temperature. These results indicate that the present method can efficiently recover noble metals from aqueous solution by changing such factors as the quantity and size of silicon powder and the hydrofluoric acid concentration and temperature of solution.

3.4.5 *Selective recovery of noble metals*

To recover and refine noble metals from industrial wastes or urban mines, the noble metal selectivity of the process is important. Figure 13 shows the recovery behavior (a) and EDS spectrum of residue (b) obtained from a mixture solution of gold, palladium, platinum, copper, nickel,

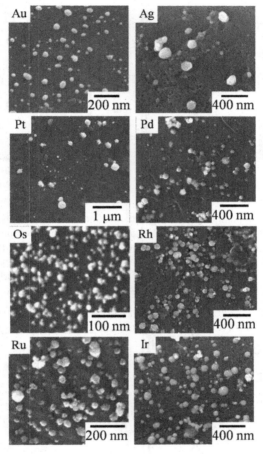

Fig. 10 SEM images of silicon powder after immersion in 1 mmol dm^{-3} metal salt solution containing 0.15 mol dm^{-3} hydrofluoric acid at 298 K. The treatment time of gold, silver, palladium, rhodium, platinum, osmium, and ruthenium is 600 s, and that of iridium is 18,000 s

cobalt, and iron chloride or chlorido-complex metal salts. Only noble metals (gold, palladium, and platinum) and copper were recovered, and no recovery, i.e. neither concentration change in the mixture solution nor characteristic X-ray of base metals (nickel, cobalt, and iron) was detected. In addition, an X-ray photoelectron spectroscopic analysis indicated the same results. The noble metal selectivity is a remarkable advantage of the present displacement deposition using silicon compared with conventional

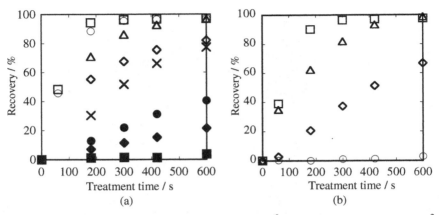

Fig. 11   Recovery of platinum. (a) Using 0.40 mol dm^{-3} (white squares), 0.20 mol dm^{-3} (white circles), 0.10 mol dm^{-3} (white triangles), 0.05 mol dm^{-3} (white diamonds), 0.025 mol dm^{-3} (crosses), 0.01 mol dm^{-3} (black circles), 0.005 mol dm^{-3} (black diamonds), and 0.001 mol dm^{-3} (black squares) silicon powder. (b) Using four kinds of silicon with different sizes: 0.016 mm powder (squares), 0.036 mm powder (triangles), 0.238 mm powder (diamonds), 10×10×0.75 mm wafer (circles)

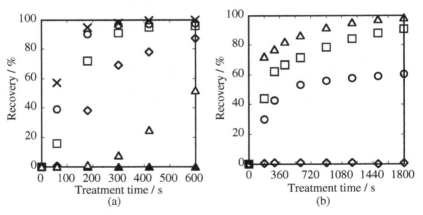

Fig. 12   Dependence of recovery of platinum (a) on the hydrofluoric acid concentration of the solution and of ruthenium (b) on the temperature of the solution. (a) The hydrofluoric acid concentration is 0.30 mol dm^{-3} (crosses), 0.15 mol dm^{-3} (circles), 0.10 mol dm^{-3} (squares), 0.05 mol dm^{-3} (diamonds), 0.03 mol dm^{-3} (white triangles), and 0.015 mol dm^{-3} (black triangles). (b) The temperature is 278 K (diamonds), 298 K (circles), 318 K (squares), and 328 K (triangles)

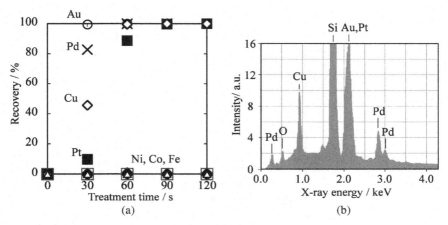

Fig. 13 Recovery of metals (a) and EDS spectrum of residue (b) obtained from mixture solution of several metal salts. Following marks represent the metal. Gold: white circles; palladium: crosses; platinum: black squares; copper: white diamonds; nickel: black diamonds; cobalt: white triangles; iron: white squares. (b) After treatment for 300 s

cementation methods using zinc or aluminum. This selectivity and potential of silicon energy band and metal deposition (Fig. 5) suggest that the electroless displacement deposition of metal proceeds by hole injection into the silicon valence band by metal deposition and silicon dissolution with the injected positive holes and hydrofluoric acid. The potential of silicon dissolution in hydrofluoric acid solution (Eqs. (43) and (44) in Fig. 5) is much more negative than that of base metal deposition such as nickel, cobalt, and iron (standard electrode potential: -0.257, -0.277, and -0.44 V, respectively). It is explained by the high stability of hydrogen termination and/or low catalytic activity for hydrogen evolution of the silicon surface. Even noble metal-deposited silicon cannot be etched in the hydrofluoric acid solution under the absence of oxidizing agent and photo irradiation except for the palladium case.

Several methods for separation and recovery of noble metals are considered: (i) separation during leaching, (ii) selective recovery using different recovery rate, and (iii) dissolving after recovery. Figure 14 shows the result of the method using the present displacement deposition on silicon powder. It is possible to selectively recover gold from mixture solution of gold and platinum ions using large difference in those velocity of recovery.

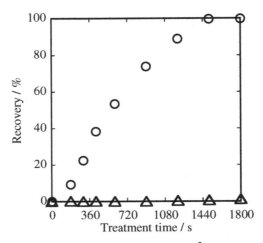

Fig. 14 Selective recovery of gold using 0.20 mol dm^{-3} silicon powder from 1 mmol dm^{-3} HAuCl$_4$ and H$_2$PtCl$_6$ mixture solution including 15 mmol dm^{-3} hydrofluoric acid. Gold: circles; platinum: triangles

The difference of recovery rate between noble metals is expected to be effective in selective recovering of specific noble metal such as gold from mixture solution of various noble and base metal ions.

3.4.6 *Recovery of gold from aqua regia solution* [81]

The influence of aqua regia in the present method was examined. Aqua regia is often used for the elution of noble metals. We used a diluted aqua regia, which is a mixed solution of 13.1 mol dm^{-3} nitric acid (HNO$_3$), 11.3 mol dm^{-3} hydrochloric acid (HCl), and water (1:4:15 in volume), including 1 mmol dm^{-3} hydrogen tetrachloroaurate(III) (HAuCl$_4$) and 0.15 mol dm^{-3} hydrofluoric acid. Figure 15 shows the recovery of gold from the solution. 100% of gold was recovered including the aqua regia by the treatment for 60 s. However, its recovery was reduced to 5% by extending the treatment time to 70,200 s. This means that gold was deposited on the silicon at the early stage of the treatment and dissolved in the solution. Recovery was increased again by adding 0.15 mol dm^{-3} of hydrofluoric acid to the solution after the recovery reduction. The reactions in the solution including aqua regia are competitive reactions of the electroless displacement deposition of gold (Eqs. (45) and (43) and/or (44)) and its

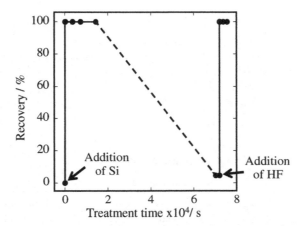

Fig. 15   Recovery of gold from HAuCl$_4$ solution including aqua regia and hydrofluoric acid by adding silicon at beginning and adding extra hydrofluoric acid at 72,000 s of treatment time [81]

chemical dissolution is expressed by the following equation:

$$Au + NOCl + Cl_2 + HCl \rightarrow HAuCl_4 + NO. \qquad (47)$$

Since the noble metal was recovered in a short treatment period, the reaction rate of the electroless displacement deposition (Eqs. (45) and (43) and/or (44)) is much higher than that of the metal dissolution by aqua regia (Eq. (47)). In a longer treatment period, the deposited gold was only dissolved by the reaction of the metal dissolution (Eq. (47)) because hydrofluoric acid was consumed by the chemical dissolution of silicon with nitrosyl chloride (NOCl) and/or nitric acid (HNO$_3$). Therefore, the recovery increased again by adding hydrofluoric acid. These results show that the recovery of noble metals from an aqueous solution including aqua regia is possible by controlling the treatment time and the concentrations of hydrofluoric acid and aqua regia. By a preliminary experiment, we successfully recovered gold from a solution that was prepared by dissolving gold foil with aqua regia.

3.4.7   *Recovery of gold using low-toxic chemicals*

The toxicity of hydrofluoric acid can be the biggest disadvantage of the present method. Ammonium fluoride is a chemical including fluoride ions

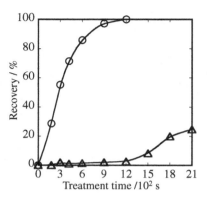

Fig. 16   Recovery of gold (circles) and platinum (triangles) from 1 mmol dm^{-3} metal–salt solution including 0.15 mol dm^{-3} ammonium fluoride and 0.2 mol dm^{-3} silicon powder

Table 4   Solution composition for electroplating of gold

$Na_3Au(SO_3)_2$	0.06 mol dm^{-3}
Na_2SO_3	0.6 mol dm^{-3}
H_3PO_4	0.3 mol dm^{-3}
pH = 0.8 (with NaOH)	

and has much lower toxicity than hydrofluoric acid. Figure 16 shows the recovery of gold and platinum using ammonium fluoride as a function of treatment time. Both, especially gold, were successfully recovered with ammonium fluoride. The large difference in the rate between gold and platinum is expected to be useful for refining. In addition, it can use sodium fluoride instead of hydrofluoric acid.

3.4.8   *Recovery of gold from secondary source of plating solution and WEEE*

As typical secondary sources of gold, we tried to recover gold from the waste solution of electroplating and waste IC chips of real waste electronics. Table 4 lists the composition of gold electrodeposition solution before using. After using the solution for gold deposition in our laboratory, the gold concentration of the solution was 10.01 g dm^{-3}, and the solution was used as the waste solution for recovery. 2.0 mol dm^{-3} of silicon powder was added in the mixture solution of 1.00 mol dm^{-3} hydrofluoric acid and the waste plating solution. After stirring for 20 min, the solution was filtered. The gold concentration of filtrate was 0.26 mg dm^{-3}. After the silicon of residue was dissolved in the mixture solution of hydrofluoric acid and nitric acid, pure gold powder was successfully obtained.

The noble metal recovery from waste IC chips was examined by using silicon powder. The milled powder of IC chips containing 0.1 wt% gold and 0.1 wt% silver was used. The other metals contained in the powder of IC chips were 19 wt% of copper, and small amount of nickel, iron, and aluminum. Figure 17 shows the flow from dissolution of metals to recovery. After dissolving silver and copper of the powder by treatment with 13.1 mol dm^{-3} nitric acid and filtering, the gold of residue was dissolved into aqua regia, which was a mixture solution of nitric acid and hydrochloric acid (1:4 in volume). The aqua regia including gold was diluted four times, 0.15 mol dm^{-3} of hydrofluoric acid and 0.20 mol dm^{-3} of silicon powder was added into the solution, and then the mixture solution was stirred for 5 min. The gold concentration of the solution was changed from 50 mg dm^{-3} to 0.09 mg dm^{-3} by the recovery. This change

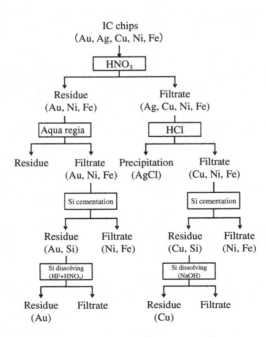

Fig. 17 Flow from dissolution of metals to recovery

corresponds to 99.8% recovery. In addition, the EDS analysis detected only gold and silicon in residue after filtering. Furthermore, dissolving silver and copper in nitric acid were subjected to recover. The concentrations of silver and copper in the nitric acid solution was 51 mg dm^{-3} and 9.2 g dm^{-3}, respectively. After precipitating, silver is removed from the solution by adding hydrochloric acid, the solution was diluted three times, and then 0.6 mol dm^{-3} hydrofluoric acid and 0.2 mol dm^{-3} silicon powder were added in the solution for recovering copper. After stirring for 15 min, the solution was filtered. The copper concentration of the filtrate was 670 mg dm^{-3} and pure copper powder was obtained by dissolving silicon of the residue into a sodium hydroxide solution. It is expected that the recovery of copper is improved by increasing amount of silicon powder. These results clearly indicate that the present method can recover pure gold and copper from the secondary source even under the presence of aqua regia.

3.4.9 *Recovery of gold using waste silicon powder (sawdust)*

The above-mentioned results were obtained by using commercial silicon sources. The sawdust silicon is waste of wafering or dicing of silicon in semiconductor industries. It is expected to be pure silicon powder usable for the present recovery process. Figure 18 shows XRD patterns of the two kinds of sawdust silicon and a commercial pure silicon powder. The patterns

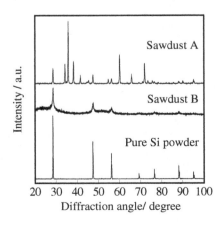

Fig. 18   XRD patterns of sawdust silicon and pure silicon powder

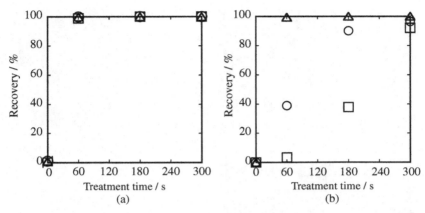

Fig. 19   Recovery of gold (a) and platinum (b) using sawdust A (triangles), sawdust B (squares), and pure silicon powder (circles)

indicate that sawdust A contained silicon and silicon carbide, and sawdust B contained only silicon. Figure 19 shows the recovery of gold (a) and platinum (b) using two kinds of sawdust silicon and the commercial pure silicon powder. Both gold and platinum were recovered by using sawdust silicon. Figure 20 shows the SEM images ((a) and (b)) and the elemental mapping of platinum (c) and silicon (d) of sawdust A after immersion in the platinum solution. The platinum particles were deposited on the silicon particles. The efficiency of the recovery using sawdust silicon resembled that using the pure silicon powder; furthermore, sawdust A gave a much higher recovery rate of platinum than that of the commercial pure silicon powder. This is caused by finer particle size of sawdust A, about 0.010 mm, than that of the commercial pure silicon powder, 0.016 mm in the mean diameter.

From the viewpoints of low cost and low environmental impact, this method is especially suitable for semiconductor industries that produce silicon powder, the sawdust silicon, and hydrofluoric acid solution as waste, and use gold and silver for electrical wiring in and between semiconductor devices. The sawdust silicon is generally disposed as industrial waste sludge. The waste hydrofluoric acid solution is generated by the cleaning and etching processes of silicon wafers. Most of the hydrofluoric acid solution is also discarded after detoxifying. Such waste semiconductor devices as IC chips and circuit boards (WEEE) are the typical ores of urban mines. The recycling cycle shown in Fig. 21 is achieved if the present

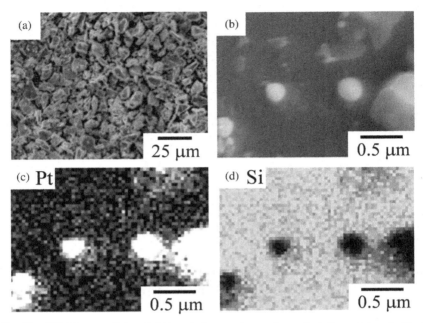

Fig. 20 SEM images ((a) and (b)) and elemental mappings of platinum (c) and silicon (d) correspond with SEM image (b) of sawdust A after being immersed in platinum solution [81]

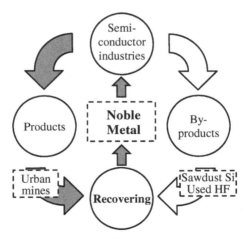

Fig. 21 Sequence of noble metal recycling using electroless displacement deposition on silicon powder for recovery [81]

method is able to recover noble metals from electronic scraps using silicon and hydrofluoric acid wastes of semiconductor industries. This sequence is expected to be low cost and low environmental impact noble metal recycling system.

3.5 Conclusions

In this chapter, the gold recovery using a displacement reaction, that is, cementation was described. The principle of electroless displacement deposition and conventional cementation processes of gold were briefly reviewed, and then the novel method to recover gold from such a secondary source as a plating solution and the WEEE using waste silicon powder were described. The cementation is traditional but an efficient and strong method to recover gold, and has been practically used in industries for over 100 years.

Recently, we developed a low-cost and efficient method to recover noble metals including gold from aqueous solution by simply adding hydrofluoric acid and silicon powder. This method has such advantages as (i) selective recovery of noble metals from a mixture solution including base metal ions, (ii) efficient recovery of gold under the presence of aqua regia, (iii) availability of selective recovery of gold from a solution including other noble metal ions, (iv) low-cost process using sawdust silicon and waste fluoride solution from semiconductor industries. We have succeeded to recover noble metals from waste IC chips which is typical WEEE. The recycling cycle of gold from WEEE using waste silicon of semiconductor industries is expected to be achieved.

Acknowledgments

The authors are grateful to Ms. Y. Ohno and Dr. N. Fukumuro for their assistance, Dong Rong Electronics Co., Ltd. for supplying sawdust silicon, and Panasonic Co. and Panasonic Eco Technology Center Co., Ltd. for lending powder of IC chips. The present work was partly supported by Exploratory Research of A-STEP from JST, Grant-in-Aid from Kawanishi Memorial ShinMaywa Education Foundation, Grant-in-Aid from Hyogo Science and Technology Association, and KAKENHI (23560875 and 26289276) from JSPS.

References

[1] Miller JD, Wan RY, Parga JR (1988) Zinc-dust cementation of silver from alkaline cyanide solutions — analysis of Merrill–Crowe plant data. *Minerals and Metallurgical Processing*, **5**(3): 170—176.

[2] Miller JD, Wan RY, Parga JR (1990) Characterization and electrochemical analysis of gold cementation from alkaline cyanide solution by suspended zinc particles. *Hydrometallurgy*, **24**(3): 373–392.

[3] Yannopoulos JC (1991) The Extractive Metallurgy of Gold, Van Nostrand Reinhold, New York, pp.185–193.

[4] Fleming CA (1992) Hydrometallurgy of precious metals recovery. *Hydrometallurgy*, **30**: 127–162.

[5] Chi G, Fuerstenau MC, Marsden JO (1997) Study of Merrill–Crowe processing. Part I: Solubility of zinc in alkaline cyanide solution. *International Journal of Mineral Processing*, **49**: 171–183 (1997).

[6] Chi G, Fuerstenau MC, Marsden JO (1997) Study of Merrill–Crowe processing. Part II: Regression analysis of plant operating data. *International Journal of Mineral Processing*, **49**: 185–192.

[7] Marsden JO, House CI (2006) The chemistry of gold extraction, 2nd edition, Society for Mining, Metallurgy, and Exploration, Inc. Colorado, pp. 365–386.

[8] Nicol MJ, Schalch E, Balestra P, Hegedus H (1979) A modern study of the kinetics and mechanism of the cementation of gold. *Journal of the Southern African Institute of Mining and Metallurgy*, **79**: 191–198.

[9] Oo MT, Tran T (1991) The effect of lead on the cementation of gold by zinc. *Hydrometallurgy*, **26**: 61–74.

[10] Hsu YJ and Tran T (1996) Selective removal of gold from copper-gold cyanide liquous by cementation using zinc. *Minerals Engineering*, **9**: 1–13.

[11] Arima H, Fujita T, W.–T. Yen (2002) Gold cementation from ammonium thiosulphate solution by zinc, copper, and aluminum powders. *Materials Transactions*, **43**: 485–493.

[12] Navarro P, Alvarez R, Vargas C, Alguacil FJ (2004) On the use of zinc for gold cementation from ammoniacal thiosulphate solutions. *Minerals Engineering*, **17**: 825–831.

[13] Cui J, Zhang L (2008) Metallurgical recovery of metals from electronic waste: A review. *Journal of Hazardous Materials*, **158**: 228–256.

[14] Karavasteva M (2010) Kinetics and deposit morphology of gold cemented on magnesium, aluminum, zinc, iron and copper from ammonium thiosulfate-ammonia solutions. *Hydrometallurgy*, **104**: 119–122.

[15] Umeda H, Sasaki A, Takahashi K, Haga K, Takasaki Y, Shibayama A (2011) Recovery and concentration of precious metals from strong acidic wastewater. *Materials Transactions*, **52**: 1462–1470.

[16] Martinez GVF, Torres JRP, García JLV, Munive GCT, Zamarripa GG (2012) Kinetic aspects of gold and silver recovery in cementation with zinc power and electrocoagulation iron process. *Advances in Chemical Engineering and Science*, **2**: 342–349.

[17] Syed S (2012) Recovery of gold from secondary sources — A review. *Hydrometallurgy*, **115–116**: 30–51.

[18] Wang Z, Chen D, Chen L (2007) Application of fluoride to enhance aluminum cementation of gold from acidic thiocyanate solution. *Hydrometallurgy*, **89**: 196–206.

[19] Nguyen HH, Tran T, Wong PLM (1997) A kinetic study of the cementation of gold from cyanide solutions onto copper. *Hydrometallurgy*, **46**: 55–69.

[20] Guerra E, Dreisinger DB (1999) A study of the factors affecting copper cementation of gold from ammoniacal thiosulphate solution. *Hydrometallurgy*, **51**: 155–172.

[21] Hiskey JB, Lee J (2003) Kinetics of gold cementation on copper in ammoniacal thiosulphate solutions. *Hydrometallurgy*, **69**: 45–56.

[22] Grosse AC, Dicinoski GW, Shaw MJ, Haddad PR (2003) Leaching and recovery of gold using ammoniacal thiosulfate leach liquors (a review). *Hydrometallurgy*, **69**: 1–21.

[23] Choo WL, Jeffrey MI (2004) An electrochemical study of copper cementation of gold(I) thiosulphate. *Hydrometallurgy*, **71**: 351–362.

[24] Kenna CC, Ritche IM, Singh P (1990) The cementation of gold by iron from cyanide solutions. *Hydrometallurgy*, **23**: 263–279.

[25] Zhang HG, Doyle JA, Kenna CC, La Brooy SA, Hefter GT, Richie IM (1996) A kinetic and electrochemical study of the cementation of gold onto mild steel from acidic thiourea solutions. *Electrochimica Acta*, **41**: 389–395.

[26] Wang Z, Chen D, Chen L (2007) Gold cementation from thiocyanate solutions by iron powder. *Minerals Engineering*, **20**: 581–590.

[27] Wang Z, Li Y, Ye C (2011) The effect of tri-sodium citrate on the cementation of gold from ferric/thiourea solutions. *Hydrometallurgy*, **110**: 128–132.

[28] Anacleto AL, Carvalho JR (1996) Mercury cementation from chloride solutions using iron, zinc and aluminium. *Minerals Engineering*, **9**: 385–397.

[29] Dönmez B, Sevim F, Saraç H (1999) A kinetic study of the cementation of copper from sulphate solutions onto a rotating aluminum disc. *Hydrometallurgy*, **53**: 145–154.

[30] Sulka GD, Jaskuła M (2003) Study of the kinetics of silver ions cementation onto copper from sulphuric acid solution. *Hydrometallurgy*, **70**: 185–196.

[31] Mubarak AA, El–Shazly AH, Konsowa AH (2004) Recovery of copper from industrial waste solution by cementation on reciprocating horizontal perforated zinc disc. *Desalination*, **167**: 127–133.

[32] Sulka GD, Jaskuła M (2004) Study of the mechanism of silver ions cementation onto copper from acidic sulphate solutions and the morphology of the silver deposit. *Hydrometallurgy*, **72**: 93–110.

[33] Fouad OA, Abdel Basir SM (2005) Cementation-induced recovery of self-assembled ultrafine copper powders from spent etching solutions of printed circuit boards. *Powder Technology*, **159**: 127–134.

[34] Karavasteva M, Kinetics and deposit morphology of copper cementation onto zinc, iron and aluminium. *Hydrometallurgy*, **76**: 149–152.

[35] Sulka GD, Jaskuła M (2005) Influence of the sulphuric acid concentration on the kinetics and mechanism of silver ion cementation on copper. *Hydrometallurgy*, **77**: 131–137.

[36] Orhan G (2005) Leaching and cementation of heavy metals from electric arc furnace dust in alkaline medium. *Hydrometallurgy*, **78**: 236–245.

[37] Demirkıran N, Ekmekyapar A, Künkül A, Baysar A (2007) A kinetic study of copper cementation with zinc in aqueous solutions. *International Journal of Mineral Processing*, **82**: 80–85.

[38] Gouvea LR, Morais CA (2007) Recovery of zinc and cadmium from industrial waste by leaching/cenentation. *Minerals Engineering*, **20**: 956–958.

[39] Gros F, Baup S, Aurousseau M (2008) Intensified recovery of copper in solution: Cementation onto iron in fixed or fluidized bed under electromagnetic field. *Chemical Engineering and Processing*, **47**: 295–302.

[40] Farahmand F, Moradkhani D, Safarzadeh MS, Rashchi F (2009) Optimization and kinetics of the cementation of lead with aluminum powder. *Hydrometallurgy*, **98**: 81–85.

[41] Aktas S (2011) Rhodium recovery from rhodium-containing waste rinsing water via cementation using zinc powder. *Hydrometallurgy*, **106**: 71–75.

[42] Gros F, Baup S, Aurousseau M (2011) Copper cementation on zinc and iron mixtures: Part 1: Results on rotating disc electrode. *Hydrometallurgy*, **106**: 127–133.

[43] Gros F, Baup S, Aurousseau M (2011) Copper cementation on zinc and iron mixtures: Part 2: Fluidized bed configuration. *Hydrometallurgy*, **106**: 119–126.

[44] Ahmed IM, El-Nadi YA, Daoud JA (2011) Cementation of copper from spent copper-pickle sulfate solution by zinc ash. *Hydrometallurgy*, **110**: 62–66.

[45] Aktas S (2012) Cementation of rhodium from waste chloride solutions using copper powder. *International Journal of Mineral Processing*, **114–117**: 100–105.

[46] Kuntyi OI, Zozla GI, Bukliv RL, Korniy SA (2013) Silver cementation from thiocyanate solutions by magnesium. *Canadian Institute of Mining, Metallurgy and Petroleum*, **52**(1): 2–6.

[47] Shibata J (1997) Cementation. *Shigen-to-Sozai*, **113**: 948–951.

[48] DIRECTIVE 2012/19/EU OF THE EUROPEAN PARLIAMENT AND OF THE COUNCIL of 4 July 2012 on waste electrical and electronic equipment (WEEE), Official Journal of the European Union, (2012).

[49] Nagahara LA, Ohmori T, Hashimoto K, Fujishima A (1993) Effects of HF solution in the electroless deposition process on silicon surfaces. *Journal of Vaccum Science & Technology*, **11**: A763–767.

[50] Morinaga H, Suyama M, Ohmi T (1994) Mechanism of metallic particle growth and metal-induced pitting on Si wafer surface in wet chemical processing. *Journal of the Electrochemical Society*, **141**: 2834–2841.

[51] Homma T, Wade CP, Chidsey CED (1998) Nucleation of trace copper on the H–Si(111) surface in aqueous fluoride solutions. *The Journal of Physical Chemistry B*, **102**: 7919–7923.

[52] Gorostiza P, Allongue P, Diaz R, Morante JR, Sanz F (2003) Electrochemical characterization of the open-circuit deposition of platinum on silicon from fluoride solutions. *The Journal of Physical Chemistry B*, **107**: 6454–6461.

[53] Harraz FA, Tsuboi T, Sasano J, Sakka T, Ogata YH (2003) Different behavior in immersion plating of nickel on porous silicon from acidic and alkaline fuluoride. *Journal of the Electrochemical Society*, **150**: C277–284.

[54] Ye W, Chang Y, Ma C, Jia B, Cao G, Wang C (2007) Electrochemical investigation of the surface energy: Effect of the HF concentration on electroless silver deposition onto p-Si(111). *Applied Surface Science*, **253**: 3419–3424.

[55] Yae S, Nasu N, Matsumoto K, Hagihara T, Fukumuro N, Matsuda H (2007) Nucleation behavior in electroless displacement deposition of metals on silicon from hydrofluoric acid solutions. *Electrochimica Acta*, **53**: 35–41.

[56] Yae S, Fukumuro N, Matsuda H (2008) Electrochemical Deposition of Metal Nanoparticles on Silicon. In Progress in Nanoparticles Research, C.T. Frisiras (ed.), Nova Science Publishers, Inc., New York. pp. 117–135.

[57] Takami K, Yae S, Yamagishi K, Fukumuro N, Matsuda H (2011) Influence of Si surface condition on electroless displacement deposition of Pt particles. *ECS Transactions*, **33**(21), 17–24.

[58] Ego T, Hagihara T, Morii Y, Fukumuro N, Yae S, Matsuda H (2013) AFM analysis for initial stage of electroless displacement deposition of silver on silicon surface. *ECS Transactions*, **50**(52): 143–153.

[59] Miyake H, Ye S, Osawa M (2002) Electroless deposition of gold thin films on silicon for surface-enhanced infrared spectroelectrochemistry. *Electrochemistry Communications*, **4**: 973–977.

[60] Miki A, Ye S, Osawa M (2002) Surface-enhanced IR absorption on platinum nanoparticles: An application to real-time monitoring of electrocatalytic reactions. *Chemical Communications*, **14**: 1500–1501.

[61] Wang H, Yan Y, Huo S, Cai W, Xu Q, Osawa M (2007) Seeded growth fabrication of Cu-on-Si electrodes for in situ ATR-SEIRAS applications. *Electrochimica Acta*, **52**: 5950–5957.

[62] Gorostiza P, Servat J, Sanz F, Morante JR (1996) The role of defects in electroless metal deposition on silicon (100). *Institute of Physics Conference Series*, **149**: 293–299.

[63] Chemla M, Homma T, Bertagna V, Erre R, Kubo N, Osaka T (2003) The role of defects in electroless metal deposition on silicon(111). *Journal of Electronalytical Chemistry*, **559**: 111–123.

[64] Fujiwara R, Hagihara T, Matsuda T, Fukumuro N, Yae S, Hitoshi Matsuda (2012) Influence of argon-plasma etching of single-crystalline silicon on electroless displacement deposition of metal particles. *Journal of The Surface Finishing Society of Japan*, **63**: 581–584.

[65] Yae S, Kawamoto Y, Tanaka H, Fukumuro N, Matsuda H (2003) Formation of porous silicon by metal particle enhanced chemical etching in HF solution and its application for efficient solar cells. *Electrochemistry Communications*, **5**: 632–636.

[66] Peng K, Zhang M, Lu A, Wong N, Zhang R, Lee S (2007) Ordered silicon nanowire arrays via nanosphere lithography and metal-induced etching. *Applied Physics Letters*, **90**: 163123.

[67] Tsujino K, Matsumura M (2007) Morphology of nanoholes formed in silicon by wet etching in solutions containing HF and H_2O_2 at different concentrations using silver nanoparticles as catalysts. *Electrochimica Acta*, **53**: 28–34.

[68] Yae S, Tashiro M, Abe M, Fukumuro N, Matsuda H (2010) High catalytic activity of palladium for metal-enhanced HF etching of Silicon. *Journal of the Electrochemical Society*, **157**: D90–93.

[69] Yae S, Morii Y, Fukumuro N, Matsuda H (2012) Catalytic activity of noble metals for metal-assisted chemical etching of silicon. *Nanoscale Research Letters*, **7**: 352–356.

[70] Yae S, Fukumuro N, Matsuda H (2009) Porous silicon formation by metal particle enhanced HF etching. In Electroanalytical Chemistry Research Trends, K. Hayashi, Editor, Nova Science Publishers, Inc., New York. pp. 107–126.

[71] Huang Z, Geyer N, Werner P, Boora J, Gösele U (2011) Metal-assisted chemical etching of silicon: A review. *Advanced Materials*, **23**: 285–308.

[72] Lévy-Clément C, (2013) Applications of porous silicon to multicrystalline silicon solar cells: State of the art. *ECS Transactions*, **50**(37): 167–180.

[73] Yae S, Kobayashi T, Kawagishi T, Fukumuro N, Matsuda H (2006) Antireflective porous layer formation on multicrystalline silicon by metal particle enhanced HF etching. *Solar Energy*, **80**: 701–706.

[74] Shinji Yae, Tsutomu Kobayashi, Makoto Abe, Noriaki Nasu, Naoki Fukumuro, Shunsuke Ogawa, Norimitsu Yoshida, Shuichi Nonomura, Yoshihiro Nakato, Hitoshi Matsuda (2007) Solar to chemical conversion using metal nanoparticle modified microcrystalline silicon thin film photoelectrode. *Solar Energy Materials and Solar Cells*, **91**: 224–229.

[75] Yae S (2011) Solar to Chemical Conversion Using Metal Nanoparticle Modified Low-Cost Silicon Photoelectrode. In Solar Cells — New Aspects and Solutions, L.A. Kosyachenko, Editor, InTech, Rijeka. pp. 231–254.

[76] Dubin VM (1992) Electroless Ni-P deposition on silicon with Pd activation. *Journal of the Electrochemical Society*, **139**: 1289–1294.

[77] Yae S, Sakabe K, Fukumuro N, Sakamoto S, Matsuda H (2011) Surface-activation process for electroless deposition of adhesive metal (Ni-b, Cu) films on Si substrates using catalytic nanoanchors. *Journal of the Electrochemical Society*, **158**: D573–577.

[78] Yae S, Enomoto M, Atsushiba H, Hasegawa A, Okayama C, Fukumuro N, Sakamoto S, Matsuda H (2013) Electroless metallization of silicon using metal nanoparticles as catalysts and binding-points electrodeposition and semiconductor metallization. *ECS Transactions*, **53**(6): 99–103.

[79] Yae S, Fukumuro N, Matsuda H (2013) Metal Nanorods in Silicon: Electroless Preparation and Application for Adhesive Film Formation. Advances in Nanotechnology. Volume 10, Z. Bartul and J. Trenor (eds.), NovaScience Publishers, Inc., New York. pp. 319–336.

[80] Fukuda K, Yae S, Fukumuro N, Sakamoto S, Matsuda H (2013) Noble metal recovering by electroless displacement deposition on silicon powder. *ECS Transactions*, **53**(19): 69–76.

[81] Fukuda K, Fukumuro N, Sakamoto S, Yae S (2014) Electroless displacement deposition of noble metal on silicon powder for recovering from urban mines. *ECS Transactions*, **61**(10): 1–7.

[82] Fukuda K, Fukumuro N, Yae S (2015) Noble metal recovery using electroless displacement deposition onto silicon. *Journal of The Surface Finishing Society of Japan*, **66**: 91–93.

[83] Paunovic M, Schlesinger M (2006) Fundamentals of electrochemical deposition Second edition, John Wiley & Sons, Inc., Hoboken, New Jersey, pp. 169–175.

[84] Bard AJ, Parsons R, Jordan J (1985) Standard Potentials in Aqueous Solution, Marcel Dekker, Inc., New York and Basel.

[85] Fujitani M, Hinogami R, Jia JG, Ishida M, Morisawa K, Yae S, Nakato Y (1997) Modulation of flat-band potential and increase in photovoltage for n-Si electrodes by formation of halogen atom terminated surface bonds. *Chemistry Letters*, **26**: 1041–1042.

Chapter 4
Adsorption of Gold on Granular Activated Carbons and New Sources of Renewable and Eco-Friendly Activated Carbons

Gerrard Eddy Jai Poinern, Shashi Sharma, and Derek Fawcett

4.1 Gold, a Historical Perspective

Gold is a metal that has fired the imagination of humanity for thousands of years and even today its allure continues to captivate the human mind. With a crustal abundance of only around 3–5 ppb, gold's scarcity has given the precious metal a unique status among many past and present civilizations. Its opaque, bright yellow color, and metallic lustre defines its physical beauty. The social structure and cultural atmosphere of many early civilizations valued gold so highly that they incorporated the ductile and malleable metal into lavish jewelry and elaborated ornaments for both decoration and religious purposes. Importantly, gold is chemically stable and not subject to atmospheric oxidation, and many of the artefacts that have survived from ancient civilizations still retain their physical beauty and lustre. Many civilizations, past and present, have used gold for the long-term building of wealth. Thus, ensuring the demand for this scarce metal is always high. Over the centuries, acquiring and building a gold

G.E.J. Poinern*, S. Sharma and D. Fawcett
Department Physics and Nanotechnology,
Murdoch University, Australia
e-mail: *g.poinern@murdoch.edu.au

reserve was a secure method for a country to develop stable, accessible, and reliable store of national wealth that could be used to offset the effects of inflation, economic crises, and geopolitical instability. Today, gold reserves held by many governments, central banks, and investment funds are used as a monetary base for supporting their international commerce and transactions. Furthermore, a country's solvency in many cases is equated to its gold reserves.

4.2 Properties of Gold

The chemical symbol for gold is Au and is derived from *aurum*, the Latin word for gold. The pronounced yellow color and metallic lustre of gold clearly distinguishes it from other pure metals that usually range in color from silvery white to various shades of grey. The unique and characteristic yellow color is due to the electronic structure of the gold atom that readily absorbs electromagnetic radiation wavelengths less than 560 nm but reflects wavelengths greater than 560 nm. Gold has a face-centered cubic crystalline structure and at the atomic level has an electron configuration [Xe] $4f^{14}5d^{10}6s^1$. The metal's color results from the oscillations of the loosely bound valence electrons orbiting the gold atom [1, 2]. Gold with an atomic number of 79 and an atomic weight of 197.0 has a relatively high density of 19.3 g cm^{-3} which makes it one of the few higher atomic number elements occurring naturally in its elemental form [3]. In its natural form, gold usually contains small amount of silver as well as those gold–silver alloys with a silver content greater than 20% are called electrum. Gold can also form other materials with minerals such as sulfides and tellurides. Gold's high density compared to solid rock allowed it to be recovered from alluvial placers using techniques such as panning and sluicing. Furthermore, gold's chemical stability and resistance to atmospheric oxidation allowed the metal to be considered as noble. Combined with various advantageous properties such as high ductility, high malleability, and a melting temperature of 1064°C, gold has been used in the manufacture of coinage, jewelry, and ornaments for thousands of years. However, the disadvantage of pure gold [24 carat] is its softness that makes it unsuitable for some forms of jewelry and ornamentation. Alloying gold with other metals can significantly change properties such as color, melting point, hardness, strength, and malleability. For example, when high percentage of copper is added to gold, its color becomes redder. The addition of small quantities of iron to gold changes its color to blue and the addition of aluminium transforms the characteristic yellow color to a purplish appearance. These alloys are rarely used since their properties make them difficult to shape and manufacture into products. This is of particular importance since around 50% of the world's consumption of new gold is used in the creation of jewelry, while investment in gold bullion by governments, central banks, and investment funds consumes 40%, and the remaining 10% is used in various industrial applications [4]. The industrial application predominantly involves depositing a thin layer of gold onto components to ensure good electrical conductivity and corrosion free electrical connectors in computers, electronic and electrical devices. Importantly, the electrical resistivity values for both silver (16 $n\Omega$.m) and copper (17 $n\Omega$.m) are superior to gold (22.1 $n\Omega$.m) for electronic connectors, but unlike gold, both silver and copper are subject to atmospheric corrosion [5]. Gold has also been used as a catalyst since its light-off temperatures (temperature when the catalyst becomes functional) are between 200 K and 350 K. This temperature range is significantly lower than the 400–800 K range for the optimal performance for a platinum catalyst [6]. The medicinal use of gold compounds has been known for centuries and was believed to be beneficial for promoting good health. Today, there is even a branch of medicine known as chrysotherapy that uses primarily gold in medicinal compounds in a number of medical protocols. Gold(I) thiol compounds have been used in the treatment of certain diseases such as rheumatoid arthritis and autoimmune diseases [1]. And recently, gold nanoparticles have been used as carriers to deliver double-stranded DNA in gene-gun technology [7]. While other studies have shown that gold nanoparticles can passively cross the cancer cell barrier and accumulate in tumors, their good optical and chemical properties can be used advantageously in thermal treatment therapies to kill those tumors [8,9].

4.3 Sources of Gold Ores

The gold present in the Earth today is believed to have come from two sources. The first is *via* a supernova event that occurred when two neutron stars had collided in space. The resulting explosion created various nuclear synthesis processes that formed a wide range of heavy elements. These expelled heavy elements condensed and then formed the solar system. During this period, the heavy elements falling onto the molten Earth ended up sinking to the core [10]. The second source comes from the bombardment of gold-loaded asteroid impacts around 4 billion years ago that enriched the Earth's crust and upper mantle [11]. The asteroid bombardment model explains why gold is widely distributed throughout the Earth's crust.

Gold ores can be formed within the Earth (Endogenetic) *via* hydrothermal processes. The process is initiated by igneous intrusions heating highly mineralized ground water. The heat enables the water to dissolve and absorb metals from the surrounding rock formations [12]. The heated, pressurized, and ore-rich waters flow via fractures in the overlying rock formations. The heat loss is experienced by the water flows as they travel through the surface formations resulting in precipitation and the formation of solid ore in fractures. These ore filled fractures or hydrothermal deposits can produce vein, and lode deposits consists of microscopic particles of elemental gold embedded within quartz, calcite, and other minerals [13]. Gold ores can also be formed at the Earth's surface (Exogenetic) due to the effects of weathering from either or both wind and water erosion of ore-rich hard rock deposits [14]. The eroded ores eventually end up in alluvial deposits called placer deposits located in river systems and coastal regions. Placer deposits can consist of free flakes, grains and due to the action of water, flakes and grains can be welded to form larger nuggets [15]. A stream placer is formed when a water flow (stream or river) carrying minerals starts to slow down. As water velocity decreases, heavy minerals drop out first and the lighter minerals continue to fall out as the velocity decreases. Because of gold's much higher density ($19.32 \, \mathrm{g \, cm^{-3}}$) compared to most rock materials ($\sim 2.7 \, \mathrm{g \, cm^{-3}}$), concentrating and extracting can be achieved using straightforward techniques such as panning and sluicing [16]. Stream placer deposits were the first to be mined in ancient times, where the river sands could be gently washed away with water leaving behind the heavier gold particles. The particles were then refined by smelting before being used in the manufacture of jewelry and ornaments.

4.4 Gold Extraction and Recovery Processes

4.4.1 *An overview of gold extraction processes*

The average concentration of naturally occurring gold in the Earth's crust is around 5 ppb with many low grade ore deposits as low as 3–10 ppm. The naturally low concentrations of gold mean that some enhancement of concentration is needed to achieve an economically viable mine development. Therefore, the nature of the ore deposit ultimately determines the type of mining operation and mineral processing techniques needed to separate the gold from the bulk ore. This can be done using natural gravity concentration methods or by leaching processes. Historically, the early miners found gold ore in stream placer deposits located around streams and rivers. In this case, gold was concentrated using gravity-based techniques such as panning. Panning involve placing an amount of ore into the pan and then adding an appropriate amount of water. The pan would then be moved around in a horizontally circular motion. During this motion, the denser gold particles would eventually sink to the bottom of the pan, while the waste, less dense sediment is washed out of the pan. Other gravity techniques using the same principle would be sluiced ore-rich sediments through jigs or over grooved surfaces that capture the denser gold while the remaining sediment flowed over.

Liberating gold from lode or vein deposits, which contained microscopic elemental gold particles embedded within a mineral matrix proved more difficult as seen in Fig. 1. Extensive crushing prior to gold extraction was required with much of the fine gold particles lost during crushing. However, around the 10th century, mixing mercury with the crushed ore to alloy with the gold particles (amalgamation) significantly improved gold recovery. For centuries, the amalgamation technique was used in refining gold, but the process is no longer used by the majority of mining companies due to mercury's highly toxic nature and the advent of more efficient alternatives. Amalgamation is still used in third-world countries due to the straightforwardness of the technique, but it is nonetheless toxic and has the constant potential to create environmental problems.

Flotation has also been successfully used in processing gold ores. In this technique, the ore is mixed with water and chemical conditioning or frothing agents to form slurry. The slurry is then subjected to intense agitation via

Fig. 1   Elemental gold embedded within a quartz matrix from a West Australian mine

the introduction of pressurized air, during which differences in the surface properties of the constituents takes place. Gold particles in the ore tend to be surrounded by air bubbles, the remainder of the ore attracts water and then sinks. The mixture of gold particles and air bubbles forms a froth that is recovered by skimming off the froth and breaking it down outside the tank, thus liberating the gold.

Prior to the 20th century, the main gold extraction route used was chlorination and several processes were in use [17]. Some examples of chlorination processes at that time were the Deetken or Mears process that used chlorine gas [18], the Munktell process that used bleaching powder and sulphuric acid, and the Black–Etard process that used a mixture of potassium permanganate, salt, and sulphuric acid. Currently, chlorine and hydrochloric acid mixtures are still utilized in some refineries to process gold and platinum group metals. The cyanide process was developed by J.S MacArthur and the Forrest brothers in 1887 to deal initially with lower grade ores in South Africa and this has become the most common process for gold extraction today [19]. The process involves the dissolution of gold (and silver if present in soluble form) from the gold bearing ore via a dilute cyanide solution. The aurocyanide-complex ions in the slurry are then exposed to granular activated carbon absorbers. The porous structure and high surface area of the activated carbon absorbs and concentrates the gold complex. Once the carbon absorber is loaded with gold ions, it is then removed from the slurry and stripped of the gold. Because of the pre-eminence of the cyanide process, it will be discussed at length in Section 4.2.

Gold ores that cannot be recovered using gravity concentration methods or cyanide leaching are usually termed as refractory ores. Therefore, a pretreatment is needed to remove much of the surrounding rock matrix and expose as many of the embedded gold particles as much as possible. In this way, the gold particles can then be leached from the ore using a conventional cyanide leaching solution. Typical pretreatments that have been used are roasting, chemical oxidation, bio-oxidation, and pressure oxidation. Roasting involves heating atmospheric air between 450°C and 750°C to burn sulphurous minerals present in the ore, thus exposing embedded gold particles. Chemical oxidation methods using nitric acid at atmospheric conditions have been used with mixed results. Bio-oxidation using sulphur consuming bacteria in water solutions have been used, but the relatively extended retention times are generally too long due to slow bacterial oxidation rates. Other techniques, using oxygen and heat in high pressure autoclaves, have been used to oxidize sulphurous mineral-based gold ores. Autoclave operations can be controlled via oxygen pressure and temperature of the chemical reactions taking place. In the case of refractory ores, the nature of the ore determines the pretreatment requirements prior to mineral processing. If extensive pretreatment of the ore is needed prior to processing, it may not be economically feasible to refine the ore, despite there being gold present in the ore.

4.4.2 Gold recovery from cyanide solutions

4.4.2.1 Introduction

Today's gold producing operations have evolved to be highly sophisticated and technically efficient processes. Even before a mining operation begins, extensive exploration, prospecting, drilling operation, and assaying are carried out to determine the feasibility and economic viability of operating the gold mine over its expected life. When the mine is operational, gold bearing ore is excavated from either open cut pits or from underground shaft operations. Prior to hydrometallurgical processing, the ore undergoes mechanical crushing. The smaller particles are then mixed with water and lime before more grinding. The finer particles are milled again in a ball mill until the bulk of the product (80%) is typically less than 70 microns. Hydrometallurgical processing of the ore consists of three major steps: (1) ore leaching, (2) solid–liquid separation, and (3) purification and concentration [20]. Furthermore, each of the steps used in ore processing is usually optimized for a particular gold ore type. After processing, the gold undergoes refining and then smelting into bullions.

4.4.2.2   Cyanidation of gold ores

Oxidation states of gold include gold(I), gold(II), gold(III), and gold(V). But, it is gold(I) (aurous compounds) and gold(III) (auric compounds) that form useful stable complex compounds [1]. Gold dissolution can be achieved with the assistance of oxidizing agents and complexes. This is why all hydrometallurgical processes include a leaching stage to transfer gold from the ore into solution as an intermediate product. Today, the cyanidation method is the most commonly used process for leaching gold from ores globally, while both thiourea and chlorine-based processes are still used in gold operations. For more than a century, the cyanidation process has been successfully used worldwide to extract gold from both high-grade and low-grade ores. High-grade ores are processed using vat-based leaching, while low-grade ores undergo heap leaching. In the vat leaching process, a mixture of sodium cyanide and ore slurry is held in large tanks for several hours while agitators mix and circulate the slurry. Heap leaching involves piling crushed ore onto an impervious ground surface covering. Then, a dilute cyanide solution is sprayed across the top of the heaped pile. With time, the solution percolates down through the heap and leaches the gold from the ore under the action of gravity. At the base of the heap, the gold laden solution is collected for gold extraction. Once the gold is removed from the solution, it is then recycled to the heap.

The cyanidation process involves the dissolution of gold (and soluble silver if present) from the ore. During this process, metallic gold (Au^0) is oxidized and dissolved in a dilute alkaline cyanide solution. Lime is added to neutralize any acidic constituents present in the ore and the oxidant used is atmospheric oxygen. During leaching, Au^0 is oxidized at the anodic spot to form Au^1, while the oxygen is being cathodically reduced at another spot on the gold surface. In solution, the Au^1 ions form complexes with cyanide ions and are stabilized [21]. The dissolution reactions for gold was originally described by Elsner in 1849, but it was not until 1954 that Kudryk and Kellogg elucidated the electrochemical fundamental properties of this reaction [22]. Since then, numerous studies have investigated the dissolution mechanisms and passivation processes involved in aerated cyanide solutions [23,24]. There has been considerable debate over the exact mechanisms at play, however, the currently accepted mechanism for the dissolution of gold from its ores using a dilute cyanide solution (usually NaCN or KCN) is described by the reactions presented in Eqs. (1)–(3) [21]. The overall dissolution reaction is described by the Elsner equation (3).

$$2Au + 4CN^- + O_2 + 2H_2O \rightleftharpoons 2Au(CN)_2^- + 4OH^-, \qquad (1)$$

$$2Au + 4CN^- + O_2 + 2H_2O \rightleftharpoons 2Au(CN)_2^- + H_2O_2 + 2OH^-, \qquad (2)$$

$$4Au + 8CN^- + O_2 + 2H_2O \rightleftharpoons 4Au(CN)_2^- + 4OH^-. \qquad (3)$$

In this oxidation mechanism, the processes proceed in three stages as presented in Fig. 2. During the first stage, an absorbed intermediate species

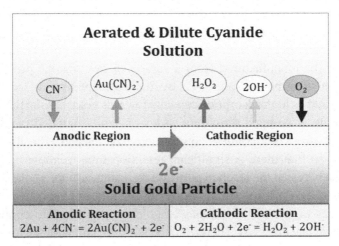

Fig. 2 Schematic representation of the electrochemical dissolution of gold into auro-cyanide ions in an aerated dilute cyanide solution

($AuCN^-_{ads}$) is formed as shown in Eq. (4). In stage two, a complex occurs between the absorbed intermediate species ($AuCN^-_{ads}$) and the free cyanide (CN^-) as seen in Eq. (5). In the third stage, a passivation layer composed of Au (III) oxide (Au_2O_3) forms over the gold surface. However, this passivation requires highly anodic potential for passivation to occur [21].

$$Au + CN^- \rightleftharpoons AuCN^-_{ads}, \tag{4}$$

$$AuCN^-_{ads} \rightarrow AuCN_{ads} + e-, \tag{5}$$

$$AuCN_{ads} + CN^- \rightleftharpoons Au(CN)_2^-. \tag{6}$$

In aerated alkaline cyanide solutions, the anodic dissolution reaction presented in Eq. (4) is accompanied simultaneously by oxygen being reduced at cathode. This reduction reaction first described by Kudryk and Kellogg is presented in Eq. (7) [22].

$$O_2 + 2H_2O + 4e^- \rightleftharpoons 4OH^-. \tag{7}$$

4.4.2.3 *Recovering gold from cyanide slurries*

4.4.2.3.1 Carbon adsorption processes

The carbon-in-pulp (CIP) is a well-known established and commonly applied process for the extraction of gold from cyanide leach slurries. Other variations of this process are carbon-in-leach (CIL) and carbon-in-column (CIC) and all processes that rely on gold being recovered from the enriched gold leached pulp by a suitable activated carbon/resin absorber. Activated carbon can be manufactured from a variety of raw organic products, but coconut shell carbon is preferred by the industry due to its strength, durability and its high adsorption capacity towards gold. During the process, leached pulp is pumped through a series of gently agitated tanks, at the same time, there is a counter flow of activated carbon. The flow rate is controlled so that there is significant retention time to allow appropriate interaction between the leached pulp and activated carbon. As the activated carbon material moves upstream the series of agitated tanks, it adsorbs and accumulates higher concentrations of gold as it comes into contact with higher grade leached pulp. Therefore, the leached pulp with the lowest gold concentration in the tanks comes into contact with the fresh or regenerated activated carbon first as seen in the schematic presented in Fig. 3 for a typical

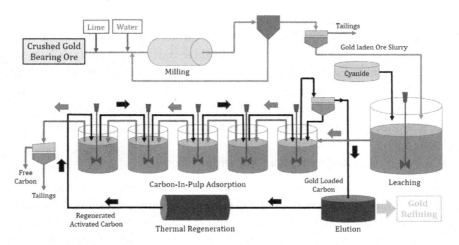

Fig. 3   Schematic of a representative CIP circuit for gold leaching and recovery

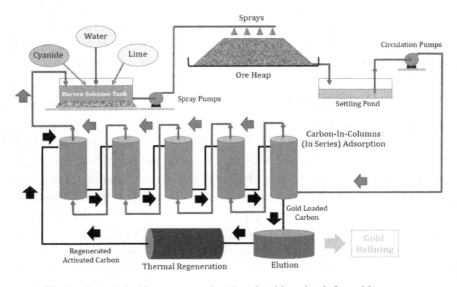

Fig. 4   Schematic of a representative Heap leaching circuit for gold recovery

CIP leaching and extraction circuit. While Fig. 4 presents a schematic diagram of a representative Heap leaching and gold extraction circuit. In Figs. 3 and 4, the black arrows indicate activated carbon flow through the circuit and the red arrows indicate the flow of leached product. Both procedures ensure efficient gold cyanide adsorption and minimize gold losses during adsorption [25]. The gold-loaded activated carbon is then separated from the leach pulp via vibrating screens that hold on to the coarse granular activated carbon. At the same time, the screens are fine enough to permit the pulp to flow through. The activated carbon is then transferred from the screens to stripping columns for desorption (elution) of gold cyanide and finally, the carbon undergoes regeneration and is recycled back into the circuit.

4.4.2.3.2 Electrowinning and refining of gold

The elution process produces a rich eluate solution that is transferred to electrowinning cells where gold and other metals are electrically deposited onto the cathodes. After deposition, the metallic material is treated with dilute sulphuric acid to dissolve metals such as zinc and copper if they are present. The remaining metallic material, usually consisting of gold and silver, is heated up to melting point to form an alloy called Dore. The Dore is then cast into moulds for assay and refining. Refining is achieved via electrolysis where silver and any platinum group materials are separated leaving pure gold. The gold can then be processed into bullion.

4.4.2.3.3 Merrill–Crowe process

The main advantage of the Merrill–Crowe process over CIP arises when there is a high silver to gold ratios. A general rule of thumb suggests that ratios greater than 4:1 favor Merrill–Crowe processing. This advantage results from the large carbon stripping and electrowinning processing required to handle large quantities of silver if present in the CIP process. The gold cyanide solution is separated from the slurry via diatomaceous earth-based filtration systems that produce a clarified solution. The gold cyanide solution then undergoes vacuum degassing to remove all oxygen contained in the solution. After oxygen removal, zinc dust is introduced to the clarified oxygen free solution. Because of cyanide's greater affinity for zinc compared to gold, the gold is precipitated along with silver in accordance with Eq. (8).

$$2KAu(CN)_2 + Zn = K_2Zn(CN)_4 + 2Au. \qquad (8)$$

The precipitate containing gold and silver is filtered from the solution. The resulting precipitate is then mixed and blended with fluxes for smelting to produce impure metallic bars. The bars are then sent for refining, however, the refining process used will be dependent on the other metals present such as copper, silver or platinum group materials.

4.4.2.3.4 Resins and ion-exchange methods

The recovery of gold from aurocyanide solutions using ion-exchange resins is not widely used as activated carbon adsorbents due to the resin cost and the cost of specialized equipment needed for resin processing [24]. Currently, the Commonwealth of Independent States (CIS-Russia) extensively uses resins for the recovery of gold from aurocyanide solutions [26]. Like the CIP/CIL processes that are based on activated carbons, there are equivalent resin processes such as resin-in-pulp (RIP) and resin-in-leach (RIL). Both activated carbon and resin-based processes are similar in principle, the major difference occurring during the elution stage. Activated carbons must first be stripped of gold and then followed by a thermal regeneration stage before the carbons can be re-introduced into the processing circuit as seen in Figs. 3 and 4. However, in contrast to activated carbons, both gold stripping and regeneration can occur simultaneously if the appropriate elution agent is used [27–29]. Crucially, the fundamental difference between activated carbons and ion-exchange resin occurs in their respective adsorption mechanism. In activated carbons, the aurocyanide complex is adsorbed between the stacked uncharged graphitic layers that form their matrix [30], while the structure of ion-exchange resins contains large numbers of desirable functional groups with similar charge that are used to attract electrostatically and concentrate oppositely charged gold ions. This feature means that resins can be specifically tailored to promote high selectivity towards the aurocyanide complex found in conventional cyanide liquors [27]. However, the competitive adsorption of unwanted anions and subsequent gold stripping from the resin has made it less competitive to current activated carbon-based extraction processes.

4.5 Activated Carbon Absorbers

4.5.1 *Activated carbon*

Carbon is one of the most abundant materials present on Earth and the arrangement of its outer electrons allows it to make four covalent bonds with other carbon atoms and other elements. This catenation property explains why there are so many different forms and structures made of carbon atoms such as chains, rings, and polygonal shapes. It is the variation of these molecular geometries that give each carbon compound its distinctive properties [31, 32]. Activated carbon is an overall generic term for a range of highly porous carbonaceous materials that cannot be exactly defined or characterized by a single physical structure or chemical analysis [33]. The lack of this common single defining physical/chemical structure or analysis stems from the wide range of organic sources such as nuts shells, coal, peat, and wood that can be processed into activated carbon. The features of activated carbons that make it a highly desirable absorber for a number of engineering applications include a highly developed and reproducible microporous structure that promotes a very large internal surface area [34]. In addition, the large internal surface area also offers a high degree of surface reactivity that promotes a considerable adsorption capacity. The versatility of activated carbons towards a number of applications such as filtration, separation, and concentration of recoverable materials results from the ability to engineer the material's microporous structure and manipulate the functional groups present on their surfaces [35, 36]. Because of these advantageous properties, activated carbons have been extensively used in food and food beverage processing, chemical and pharmaceutical processing, gas separation and purification, and also water treatment operations [37–39]. The popularity of activated carbons in these particular applications is also reflected in the actual volumes being used. For example, world consumption in 2002 was around 750,000 tonnes [40–42] and during the following decade, consumption steadily increased to around 1.2 million tonnes by 2012 with an estimated value of USD 1,913 million. Forecasting for the 2013–2019 period by industry analysts indicates that the tonnage is expected to increase annually by 10.2%, while the expected value is predicted to increase by 11.9% annually and by 2019, the estimated value will be around USD 4,180 million [43].

4.5.2 *Source materials for activated carbons*

Activated carbons can be produced from a wide range of raw organic materials derived mainly from plants. Sources also include coal, peat, lignin, wood, and vegetable wastes from food processing. In terms of a mechanically strong and durable activated carbon, the hard shell of the coconut has been extensively used by the gold industry because of its superior adsorption properties coupled with its inherent strength [43]. Other nut-shell sources include almond, hazelnut, macadamia, pecan, and walnut while fruit stones such as apricot, cherry, peach, and plum have all been studied as precursor for activated carbons [44–51]. The importance of the inherent properties of the raw material cannot be underestimated since these properties have a significant influence on the final chemical and physical properties of the resulting activated carbon [52]. Highly desirable inherent properties include (1) high microporosity, (2) intrinsic hardness, (3) high density, (4) low attrition loss, (5) long service life, and (6) low ash content. For example, woody-based activated carbons with 1 g having more than 500 m^2 of surface area make good decolorizing agents for dye-contaminated water [53]. Alternative sources of raw materials that have been used to make activated carbons include bone, sugar, and tyres, agricultural waste and waste materials from municipal or industrial waste water treatment facilities [54–60].

There are two main forms of activated carbon. Powder is the first form and it has a large external surface area with a relatively small diffusion distance inside its pores. Thus, making the rate of adsorption very high and making this product ideal for solution-based adsorption applications. The raw material for these powders is usually sawdust, which after carbonization is chemically activated. But, unlike many granular forms, powdered activated carbons are not reusable [46]. Granular activated carbons are the second form of the material. These activated carbons are known for their highly porous structure giving them very high surface areas. The porous structure is composed of micropores (pore dia. < 2 nm) and mesopores (pore dia. 2–50 nm) that promote the transport of molecules or particles within the porous matrix. This extensive array of interconnecting pore channels dramatically increases the surface area and greatly improves the adsorption capability of the activated carbon [61]. Furthermore, studies have shown the presence of chemically bonded heteroatoms located at the edges of carbon layers. The presence of heteroatoms such as oxygen, nitrogen, and hydrogen can influence the types of surface functional groups that can interact with the activated carbon [62]. Studies have shown that modifying the surface functional groups containing oxygen and nitrogen via thermal and chemical methods to promote coordination chemistry that can attract a variety of heavy metals [63–65]. Hence, granulated activated carbons have been extensively used in water purification as well as for the removal of toxic effluents from contaminated waters [66, 67].

In terms of gold recovery, Johnson patented a process that used a cyanide leach solution and activated carbon derived from wood charcoal [68]. However, a practical process for using cyanidation and granulated activated carbon to recover gold from solutions was only developed in the 1950s by Zadra *et al.* while working for the US Bureau of Mines [69]. Since then, there has been considerable interest from both the scientific community and industrial fields in investigating, understanding, and optimizing the adsorption and subsequent desorption of gold from activated carbons. Over the years, the gold mining industry has predominantly focused on two activated carbon-based technologies, namely CIP and CIL. Both technologies have enabled the industry to maximize extraction rates from ores, and in turn reduce processing and operating costs. Each process uses activated carbons derived from either peat or coconut shells, both of which have effective surface areas around 1000 m^2/g. In particular, activated carbons derived from coconut shells are strong, durable, and have a long service life, which mean that they can be regenerated many times.

4.5.3 *Carbonization and activation of raw materials*

4.5.3.1 *Introduction*

In general, the procedure for producing activated carbon is a two-step process. The first step is carbonization and its objective is to create a carbon structure with a high carbon content (ideally 80% and above). To achieve this goal, a pre-sized organic raw material is heated in an inert atmosphere. During this step, dehydration and the removal of volatiles take place in controlled temperature dependent procedure. Activation is the second step and involves developing a porous structure of molecular dimensions that dramatically increases the surface area of carbon. The final pore interconnectivity, pore volume, mean pore diameter, and surface area of the final activated carbon are all dependent on the nature of the raw material, temperature, and activation time and the type of oxidizing agent used [70, 71].

4.5.3.2 *Carbonization of raw materials*

Carbonization is performed at temperatures between 400°C and 800°C in the presence of an inert gas atmosphere on the organic raw materials. For example, commercial coconut-based carbons are usually carbonized at 500°C. The thermal treatments are designed to dehydrate, remove non-carbon species (carbon dioxide, carbon monoxide, oxygen, nitrogen, and hydrogen) and thermally decompose any volatile components [72]. Thermal treatment produces a carbonized mass with a poorly developed porous structure within the carbonized mass matrix [48,72]. Manufacturers of activated carbon generally use a pretreatment procedure using dehydrating agents such as phosphoric acid, zinc chloride, or sulphuric acid to encourage porosity during carbonization [73,74]. During thermal treatment, the agents influence the pyrolytic decomposition, restrain tar formation, reduce the formation of acetic acid and methanol, and significantly improve the carbon yield. Lower treatment temperatures generally improve the size, interconnectivity and distribution of the pore structure within the final carbonized mass. Thus, the amount and type of agent combined with the appropriate treatment temperature are important factors in determining the quality and quantity of the carbon produced. Crucially, higher amounts of pretreatment agents result in larger pore structures being formed in the carbonized mass [44,71,75]. After carbonization, the carbon atoms present in the mass are grouped into stacks of flat cross-linked aromatic sheets. The spaces between the sheets that form the pores in some cases can be filled with products of decomposition or disordered carbon if the pretreatment agents fail to impregnate the mass [71]. Because of the poorly developed porous structure and low surface areas produced during carbonization, the adsorption capacity of the resulting carbonized mass is relatively low [57]. Therefore, a second procedure (activation) is used to further develop the porous structure and amplify the adsorption properties of the carbonized mass.

4.5.3.3 *Activation and carbon structure*

Activation leads to further development of the primary porous structure including further pore creation and dramatically increases the surface area of the various internal pores structures. Activation is typically carried out between 800°C and 1100°C in an oxidizing atmosphere. Typical atmospheres can consist of flue gas (source of carbon dioxide (CO_2)), steam or atmospheric air or various combinations of these gases [40,50,76]. During activation, disordered carbon located in spaces between the aromatic sheets and more reactive regions of the carbonized structure are exposed to oxygen and undergo endothermic reactions as described by Eqs. (9) and (10). Furthermore, steam (H_2O) reacting with the carbon structure produces carbon monoxide (CO), which is then accompanied by water to gas formation as described by Eq. (11) [71,76].

$$C + H_2O = CO + H_2 \qquad \Delta H^o_{1073\,K} = 136\,kJ, \qquad (9)$$

$$C + CO_2 = 2CO \qquad \Delta H^o_{1073\,K} = 170\,kJ, \qquad (10)$$

$$CO + H_2O = CO_2 + H_2 \qquad \Delta H^o_{1073\,K} = -34\,kJ. \qquad (11)$$

The extent of activation will depend on the original organic material, type of gas or gas mixture used, the treatment temperature and overall treatment times [48]. The burnout of carbonaceous compounds and disorganized carbons between the aromatic sheets produce the final pore structure. During this time, significant pore widening occurs and produces pores with very large internal surface areas [45]. However, the burnout is not a uniform process due to difference in the crystalline properties produced during the initial carbonization process and results in variations in the size, texture, and distribution of pores. For example, when burnout is less than 50%, a micropore structure is formed. However, when burnout is between 50% and 75%, a widely varying pore structure is produced. Thus, by controlling the activation parameters, it is possible to engineer activated carbons with specific properties for particular applications [21,33,77,78].

4.5.4 *Physical and chemical properties of activated carbons*

The matrix of graphite is similar to activated carbons, but with a superior degree of order. X-ray diffraction studies have identified the presence of two types of structures occurring in activated carbons [18]. The first type is composed of regions of elementary crystallites comprising hexagonally ordered carbon atoms. In contrast, the second type is formed of disordered and cross-linked lattices of hexagonally ordered carbon atoms [79]. The

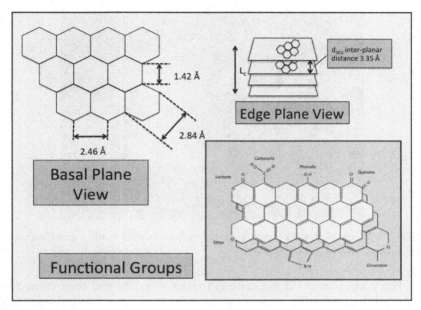

Fig. 5 Hexagonal arrangement of carbon atoms and some functional groups present in graphite and activated carbons

sizes of these elementary carbons are the direct result of the temperature used during activation. Typically, their width ranges from 2.0 nm to 2.3 nm while their heights vary 0.9–1.2 nm in height (see Fig. 5). Correspondingly, the nanometer scale crystallites present are usually around the diameter of a hexagonal arrangement of carbon atoms and a height of three layers [80].

Activated carbons are distinguished from other materials by their formidable adsorption capacity, high degree of porosity, and pore size distribution. In the case of activated carbons, Dubinin classified the size distribution of the pores into three groups [81]:

Type of Pore	Pore width (w) range
Micropores	$w < 2\,\text{nm}$
Transition pores/Mesopores	$2\,\text{nm} < w < 50\,\text{nm}$
Macropores	$w > 50\,\text{nm}$

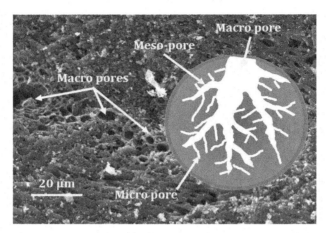

Fig. 6 SEM image of the coconut activated carbon showing surface pores and insert showing a representative pore structure

The classification of the different types of pores and their respective diameters is based on their width w [82]. The macropores contribute very little to the total surface area and do not play any significant role in promoting adsorption, but provide passage to the mesopores and the micropores as seen in Fig. 6. Both the mesopores and the micropores provide access to the interior of the carbon matrix. Importantly, the micropores formed and developed during activation are responsible for the very large surface area normally associated with activated carbons. Surface area measurements of solids are generally made using the Brunauer, Emmett, and Teller (BET) method. In this technique, the BET surface area value expresses the surface area in square meters per gram of adsorbent, covered by a monomolecular layer of nitrogen gas [83]. Generally, activated carbons have BET surface areas ranging from 500 m^2/g to 1400 m^2/g and for gold adsorption applications, it is typically around 1000 m^2/g. The mesopores make up less than 5% of the total internal surface area, while the micropores account for 95% of the total internal surface area and are the most important feature associated with the adsorptive action of the carbon [56].

The adsorption capacity of any activated carbon is not only determined by its physical porous structure, but also on its surface chemistry. The effect of surface chemistry and its influence in adsorption processes is not fully understood, but the surface chemistry of activated carbons can be associated with three main causes [18]: (1) Disruptions in the normal microcrystalline structure caused by edge and dislocation effects form residual carbon valances that can influence the adsorption of polar and polarisable species; (2) The presence of chemically bonded elements such as oxygen, nitrogen, and hydrogen in the original source material or chemical bonding between the carbonized mass and chemical species existing in the activating agents, and (3) The presence of inorganic matter such as ash components or residual activating agents that either enhance or adversely influence adsorption.

In terms of surface oxides, despite being a relatively minor component, they can have a drastic or significant influence on the overall chemical behavior of the carbon. The presence of these surface oxides can produce acid–base, redox, and catalytic properties as well as imparting a hydrophilic character to naturally hydrophobic carbon. For example, oxygen chemisorbed onto carbon can readily form C–O functional groups that can subsequently influence surface reactions, wettability, and electrical property of the activated carbon [18, 84]. Because of the effect of surface oxidation on surface chemistry, activated carbons are generally classified into two categories based on their acid–base character. The two categories are H-carbons and L-carbons, and are distinguished from each other by their respective pH values. H-carbons are characterized by their ability to adsorb H^+ ions when immersed in aqueous solutions with the result being an increase in the pH of the resulting solution, whereas L-type carbons preferentially adsorb OH^- ions present and generate higher pH values in the resulting solution [85]. H-carbons (protonated surfaces, $C–OH_2^+$) are activated at temperatures above 700°C with 1000°C being a typical thermal value. L-carbons (ionized surfaces, CO^-) are typically activated below 700°C, usually between 300°C and 400°C. Steam activated carbons are generally used for gold recovery and are typically activated at temperatures between 600°C and 700°C. And as a result, they can display both acidic and basic properties, but generally with more H-carbon type characteristics [18].

4.5.5 *Properties of activated carbons in the hydrometallurgy industry*

4.5.5.1 *Introduction*

Historically, charcoal material was first used to recover gold from ores, but the subsequent burning of the charcoal and smelting the ash to recover the gold proved to be uneconomical. However, the introduction of activated carbons into large-scale cyanide-based leaching processes by the gold industry in the 1970s made it possible to extract gold from low-grade ores and tailings economically. For optimum performance in a gold recovery operation, the selected activated carbon must have important properties such as high rates of adsorption, high gold-loading capacity together with high resistance to abrasion in the circuit [86]. These properties have made the use of activated carbon technically and economically more attractive compared to earlier gold recovering processes using cyanide leach solutions [87]. Two commercial types of activated carbons currently being used by industry are derived from peat and coconut shells. Figures 7 and 8 present scanning electron microscopy (SEM) surface images of commercial peat and coconut shell derived activated carbons. Both sets of figures clearly show the respective porous surface features of both activated carbons. In particular, factors such as high selectivity towards gold and the ability to reuse the absorbent in the extraction process has made the use of activated carbons an attractive option since the technology was introduced, thus making activated carbon-based CIP and CIL technologies

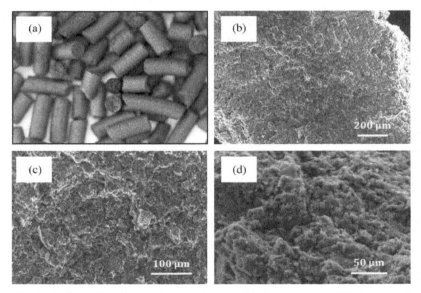

Fig. 7 Optical and SEM images showing the porous surface features of Norit peat derived active carbon

Fig. 8 Optical and SEM images showing the highly porous surface of coconut shell derived (Haycarb) activated carbon

the standard for the majority of gold mining operations throughout the world [88, 89].

4.5.5.2 *Gold adsorption mechanisms and loading on activated carbons*

Gold adsorption on activated carbons is dependent on a number of chemical and physical factors that influence both adsorption kinetics and gold-loading capacity. Initially, the adsorption rate of gold cyanide is fairly rapid and adsorption takes place on all accessible sites. Most of these adsorption sites are located in and around the macropores and possibly some of the mesopores. As the mass transport of gold cyanide species continues, adsorption kinetics steadily decreases until a pseudo-equilibrium is reached. From this point onwards, a much slower adsorption process takes place within the micropores [18]. Adsorption at this stage is much slower due to the effects of diffusion taking place along the tortuously long lengths of the micropores. Several studies have been undertaken to explain and model the complex nature of the diffusion mechanisms involved during the adsorption of gold cyanide on porous active carbons. These studies have proposed three rates of dependent diffusion mechanisms based on contact time and gold-loading capacity to explain the adsorption of gold cyanide. The three mechanisms are (1) film diffusion (mass transport), (2) pore diffusion, and (3) surface diffusion [87, 90–97]. And as discussed above, initial gold cyanide adsorption is rapid with adsorption taking place on all easily accessible sites and is explained by film diffusion (mass transport). Pseudo-equilibrium is reached when maximum adsorption capacity is achieved in the macropores and mesopores. Beyond pseudo-equilibrium, further adsorption takes place in the micropores. Film diffusion in the micropores is much slower and is more difficult due to the length and tortuosities of the micropore structures present in the activated carbon. Recent microtomography studies by Dai and Pleysier in Western Australia have confirmed the poor diffusion of gold cyanide within micropore structures. And as a result, the studies rule out the influence of pore diffusion and suggest surface diffusion is more dominant at higher gold loadings [92,97].

To clarify the position/location of gold cyanide complex (aurocyanide) adsorption sites on activated carbon surfaces, several studies have been undertaken. However, these studies have produced conflicting results and have consequently made it difficult to determine the exact adsorption mechanism. Surveying the literature reveals two possible mechanisms that can explain the adsorption of aurocyanide on the surfaces of micropores: (1) aurocyanide is adsorbed as an ion-pair of the form $M^{n+}Au(CN)_2^-$, and (2) aurocyanide is adsorbed via the decomposition of $Au(CN)_2^-$ to form $AuCN$ [98–101]. Reviewing the literature reveals that the ion-pair model is the most widely accepted mechanism and studies have confirmed that aurocyanide adsorbs onto the walls of micropore as an ion-pair $M^{n+}Au(CN)_2^-$ [102–105]. In the ion-pair mechanism, M^{n+} is a cation that is most likely potassium, sodium, or calcium [98, 102–104]. Furthermore, functional groups produced during activation were also found to be responsible for the adsorption of aurocyanide [105–107]. The ion-pair mechanism is explained by Eq. (12), where $M^{n+}[Au(CN)_2^-]_n$ term is the adsorbed gold species attaching to the walls of the micropores.

$$M^{n+} + nAu(CN)_2^- = M^{n+}[Au(CN)_2^-]_n. \tag{12}$$

Spectroscopy studies have been used to investigate the nature of aurocyanide adsorption on activated carbon surfaces. Mössbauer spectroscopy studies have shown that aurocyanide anions adsorb to activated carbons surfaces through one of the cyanide ligands [108, 109]. While both X-ray photoelectron spectroscopic (XPS) and infrared (IR) studies have shown that aurocyanide ions tend to adsorb onto the small graphitic plates found in the activated carbon structure through an intact complex ion with a partial transfer of electrons to the gold (I) ion [110, 111]. Further adsorption studies have examined the influence of graphitic components found in activated carbons. The studies used ^{14}C labeled aurocyanide and revealed that adsorption predominantly occurred on the edge planes that occur in highly orientated pyrolytic graphite (HOPG) [112]. Similar aurocyanide adsorption studies using HOPG in the presence of Ca^{2+} via *in situ* scanning tunnelling microscopy was able to locate atomic scale clusters of gold. The gold was predominantly found along graphite basal and edge planes [113]. Similar *in situ* STM studies by Poinen *et al.* in aurocyanide solutions with calcium ions have shown atomically-sized filamentous gold structures present along and on the planes of HOPG (see Fig. 9) [114].

From an operational point of view for the gold industry, carbon activity is an important factor in determining aurocyanide adsorption within the process circuit. Importantly, carbon activity governs the amount of activated carbon necessary in each tank to carry out maximum adsorption. It also determines the required contact time needed, thus determining the size of the tank and the flow rate through the tank. And overall, carbon activity is used to determine gold within the process circuit and operational plant efficiency [115]. Mathematically, carbon activity and the subsequent gold loading within a gold processing circuit can be modeled using Eqs. (13)–(17). Equations (13) and (14) are used to determine the rate constant (k) and equilibrium constant (K).

$$\ln \left[\frac{[Au]_{s,0} - B}{[Au]_s - B} \right] = k \left[\frac{KM_c}{M_s} + 1 \right] t, \tag{13}$$

$$B = \frac{M_s[Au]_{s,0} + M_c[Au]_{c,0}}{KM_c + M_s}. \tag{14}$$

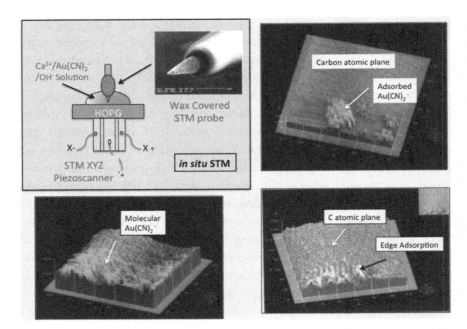

Fig. 9 *In situ* scanning tunnelling microscopy of HOPG under an alkaline aurocyanide (with added Ca^{2+} ions) solution with micrographs depicting the carbon atomic plane and molecular adsorption of the aurocyanide complex (in plane and at the edge) [113, 114]

Given

$[Au]_s$ or $[Au]_s^t$ = concentration of gold in solution at time t,

$[Au]_c$ = concentration of gold on the carbon at time t,

$[Au]_{s,0}$ = concentration of gold at t = 0,

$[Au]_{c,0}$ = concentration of gold in the carbon at t = 0,

$\Delta[Au]_s^t$ = change in gold loading onto carbon from time zero to t,

M_c = mass of the carbon,

M_s = mass of the solution,

k, k' = rate constant,

K = equilibrium constant.

Both these constants are determined from a series of batch experiments where the gold concentration is continuously changing. Both constants are then used in Eq. (15) along with a constant gold concentration to determine the rate of gold holding in each stage of the processing circuit [90, 91, 116, 117].

$$\frac{d[Au]_c}{dt} = k(K[Au]_s - [Au]_c). \tag{15}$$

Furthermore, carbon activity can also be expressed by an empirical rate constant (k') that is determined experimentally by batch testing and is presented mathematically in Eqs. (16) and (17), where n = empirical constant. This alternative approach is frequently used for comparing different types of activated carbon, fouling of activated carbon, and the influence of dewatering reagents [118–120]. Moreover, comparative studies of various kinetic models have shown that Eq. (16) is suitable for modeling experimental data over an 8 h period [121].

$$\Delta[Au]_c^t = k'[Au]_s^t t^n, \tag{16}$$

$$\log \frac{\Delta[Au]_c^t}{[Au]_s^t} = n \log t + \log k'. \tag{17}$$

4.5.5.3 *Abrasion resistance and mechanical properties of activated carbons*

To be of practical benefit in the gold extraction processes, activated carbons need to have good mechanical properties that enable them to withstand the severe abrasion conditions encountered in typical aurocyanide adsorption circuits [46]. The breakdown of activated carbons in the gold extraction process not only represents the loss of carbon itself but also translates into a significant loss in gold. The major breakdown of activated carbons occurs as a result of wet attrition that occurs in the abrasive agitated pulp environment of the adsorption circuit. Other forms of attrition occur from dry abrasion, screening operations via transport within the circuit and during regeneration [87]. A technique that has been used to give indication of the performance of an activated carbon's resistance to particle size degradation under service conditions is the ball-pan method [122]. However, this method does not measure in-service resistance to degradation, but it does establish a measurable characteristic that can be used to grade and compare various activated carbons by the industry. The method involves placing a screened and weighed sample of activated carbon into a pan (of known hardness) with a number of stainless steel balls. The pan is then subjected to a combined rotating motion and tapping action for 30 min. At the end of this period, degradation is determined by weighing the quantity of carbon retained in the sieve. Based on this type of attrition testing, activated carbons are usually conditioned to remove imperfections and weak features in the granules before being introduced into the gold processing circuit. Prior conditioning minimizes the possibility of losing gold on undersize or structurally inferior carbon particles within the circuit. Importantly, conditioning can often indicate how effective the manufacturers have been careful in preparing the activated carbon [123]. Variations in the hardness of activated carbons can result from differences in the pore structure since softer carbon results from more of the carbon matrix around the pore structure being burned away during activation. Thus, making them less dense compared to activated carbons that have not experienced excessive burnout.

4.6 Renewable and Eco-Friendly Sources of Activated Carbons

4.6.1 *Agricultural Waste as Precursors for Activated Carbon Production*

Activated carbon can be generated from a wide range of natural and synthetic solid carbonaceous precursor materials. The two major sources of precursor materials include coal and agricultural crops and by-products [124–127]. Historically, many commercial activated carbons have been made from raw materials derived from coal, petroleum by-products, lignite, and peat [128–130]. Currently, the precursor sources for commercial activated carbons are derived from two main sources: (1) coal/lignite (42%) and (2) coconut/wood (45%) [131]. However, many of these activated carbons are expensive and non-renewable, which make them economically undesirable and ecologically unsustainable. Essentially, any material or compound with a high level of carbon and a low inorganic content can be used as a raw material for the production of activated carbons [126]. However, other factors such as the cost of the raw material as feedstock, composition of the raw material, the type and extent of activation needed and supply security must also be considered [132]. In particular, the most important factor that needs to be considered prior to converting a material into activated carbon is its physical and chemical properties [133]. For example, the raw material should have a high carbon content and have in-built porosity since these properties will significantly influence the character and nature of resulting activated carbon [134]. Furthermore, the influence of processing parameters such as carbonization temperatures, pretreatment additives, treatment periods, activation agents and temperatures all have an effect on the resulting activated carbons generated [135].

In recent years, the prospect of converting agricultural by-products into activated carbons has attracted considerable interest since it offers a renewable and relatively inexpensive source of raw materials [136,137]. The carbon content of many of these by-products is lower compared to more traditional precursor materials such as coal and peat, but their volume and lower costs have attracted the attention of both scientists and the agricultural sector [138]. To the agricultural sector, it offers an opportunity to produce an economically viable and renewable product from waste by-products and offset the costs currently incurred in processing, transporting, and disposing of waste by-products. Moreover, producing activated carbon from agricultural by-products can significantly reduce the environmental impact of waste pollution. For example, a wide range of agricultural waste by-products such as bamboo [139], husks [140], stalks [141], wood [142–144], nutshells, and stones [145,146] are potentially eco-friendly and renewable precursors for activated carbon manufacture.

4.6.2 *Nutshells and fruit stones*

Millions of tonnes of waste nut shells and stones from a variety of fruits and nuts are produced by the global agricultural/food processing industry every year. Currently, the individual processors carry the cost of disposing very large quantities of waste nutshells and stones. Unfortunately, the volume of waste produced by annual crop production generally maintains this disposal cost. In particular, the cost will continue to increase with increasing crop production and the increasing costs associated with waste disposal. Any agricultural waste material, in particular, nutshells and stones from fruits are high in carbon which makes them an ideal precursor for the production of activated carbons. Using the agricultural waste from nuts and food-processed stone fruits as activated carbon precursors offers a renewable and eco-friendly resource that could effectively convert the waste into economically viable products [132]. For the last decade, studies have shown the potential of using nutshell and fruit stones as activated carbon precursors [147–149]. Recent studies have shown nutshells derived almond [150], hazelnut [151], macadamia [50, 61, 152], pistachio [153], pecan [154], and walnut shells [155, 156] from food processing industries are ideal precursors for activated carbons, while fruit stones from apricot [157–159], cherry [160], olives [161, 162], and peaches [163, 164] have all been used as precursors for the preparation of activated carbons. For example, Yalcin and Arol have investigated the adsorption of aurocyanide on activated carbons derived from hazel nutshells, peach, and apricot stones to ascertain their suitability as an alternative to activated carbon produced from coconut shells [48]. These adsorption studies found that activated carbon produced from both apricot and peach stones were comparable to those of currently used commercial products including coconut-based activated carbon. Similar studies have shown that relatively inexpensive waste products from the nut and fruit processing industries are capable of producing activated carbons high microporosity and high surface areas. However, more detailed studies are needed to examine the production, optimization, and application of activated carbon derived from precursor nutshell and fruit stone waste before this source of inexpensive agricultural waste can be commercially competitive with currently available activated carbon products in the market.

4.6.3 *Case study: Macadamia nut shells*

4.6.3.1 *Introduction*

Surveying the literature reveals that in recent years, there has been significant interest in using naturally available agricultural wastes for the production of activated carbons [136, 137]. Today, Australia uses approximately 18,000 hectares consisting of around 5.5 million trees for the cultivation of macadamia nuts along the coastal regions of northern New South Wales and Queensland. Macadamia nuts produced in these regions accounts for 40% of the global market and is valued at around AUS $150 million. Unfortunately, a consequence of producing large quantities of macadamia nuts is the serious problem of producing large quantities of waste nut shells and how to effectively dispose them [165]. In recent years, around 5 Mt of waste shells have been used as fuel to produce electricity for macadamia plant operations in the Gympie region of Queensland, thus only partially alleviating the waste problem [166]. However, falling coal prices are making nutshell-based fuel less competitive. Further complications arise because less than half of the waste can be used as fuel and increasing production levels are expected to increase annual waste material at rate of 10% [167].

Currently, the Australian gold industry imports peat and coconut shell derived activated carbons from overseas at a significant annual cost [168]. Therefore, an opportunity exists for converting waste macadamia nut shells into activated carbon and potentially reducing the operating costs of the Australian gold industry and also reduces their reliance on imported activated carbons. Importantly, macadamia nuts have a large cracking pressure (2.07 MPa or 300 psi) which indicates the inherent strength of the shell and should lead to a more abrasion resistant activated carbon suitable for CIP/CIL applications. Furthermore, adsorption studies using activated carbon derived from macadamia shells could be successfully used to remove organic contaminants from aqueous solutions and also could be used as precursor filters for molecular sieves [137,169,170]. Similar studies have shown that macadamia shell based activated carbons can adsorb a wide range of organic molecules and a range of metallic ions such as copper and gold [49, 50].

4.6.3.2 *Conversion of waste macadamia nut shells to activated carbon granules*

The conversion procedure for producing activated carbon from a suitable raw material is two-step procedure. In the present case, a similar two-step procedure was followed and in the first step, waste macadamia shells were carbonized at 500°C for 45 min in a nitrogen atmosphere to prevent oxygenation of the carbon product. Carbonization also removed both water and volatiles retained by the shells. The following activation step was designed to promote the development of an extensive porous structure of molecular dimensions that dramatically increased the effective surface area of the carbon matrix. The activation agent used was carbon dioxide (CO_2)

and produced a significant increase in the BET surface area with increasing activation temperatures up to 900°C (1104.15 m^2/g) for treatment periods of 30 min. Similar studies carried out on coconut shells using the same two-step processing procedure produced a surface area of 1210.66 m^2/g at 900°C. Indicating that the macadamia shell derived activated carbon had a comparable surface area to activated carbons derived from coconut shells. Beyond this temperature, the surface area of macadamia derived activated carbon steadily declined and resulted from the degradation of the porous structure. The study revealed that the development of the porous structure and ultimately the surface area were dependent on the properties of the initial macadamia shell, activation temperature, and activation time [50]. Figure 10(a) presents an image of waste macadamia nutshells prior to the two-step conversion process and Fig. 10(b) presents the final activated carbon product after processing.

4.6.3.3 *SEM study of activated carbon granules*

SEM studies were undertaken to investigate the formation of porosity during activation. A representative image of an unprocessed shell is presented in Fig. 9(a) and reveals a non-porous and rough surface texture. Measurements indicate shell granules in the unprocessed state generally have surface areas ranging from 0.1 m^2/g up to around 0.15 m^2/g. The subsequent carbonization procedure resulted in both water and non-carbon materials such as waxes and oils being eliminated. Also occurring was the thermal decomposition of the shell to produce a solid carbon structure with a

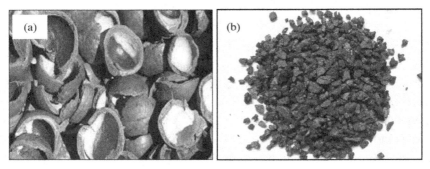

Fig. 10   (a) Waste macadamia nutshells prior to treatment and (b) resulting activated carbon produced at the end of the activation process

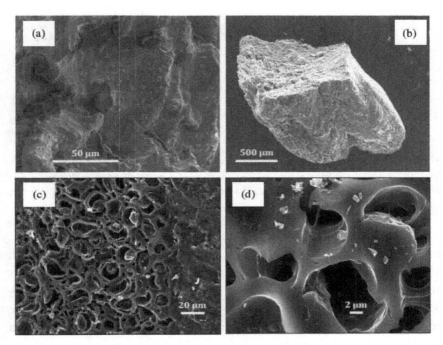

Fig. 11 Pore structuring and surface remodeling produced by the carbonization and activation steps; (a) surface of shell prior to processing; (b) representative activated carbon granule produced from a macadamia shell; (c) enlarged image of granule showing surface pore structure; (d) expanded image showing pores

rudimentary pore structure. During the following activation step, there was a significant improvement in pore development that produced highly porous activated carbon granules. Figure 11(b) presents a representative activated carbon granule produced from waste macadamia shells. The overall size of the granule is around 2 mm and surface features can be seen. The enlarged surface image presented in Fig. 11(c) reveals extensive pore development that is indicative of a highly porous material. Compared to the unprocessed shell granule presented in Fig. 11(a), we can see that carbonization and activation has created a highly porous structure with numerous surface pores providing access to inner regions of the granule as previous discussed and graphically presented in Fig. 6.

The growth in porosity corresponds to a dramatic increase in surface area and both are dependent on the activation temperature. This also

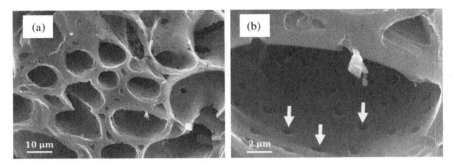

Fig. 12   (a) Cross-section view showing the extensive pore structure with macropores clearly evident and (b) enlarged view of a single pore showing an internal array of other micorpores.

translates to the adsorption capacity of the resulting activated carbon. Principally, adsorption occurs in the micropores, which generally accounts for about 95% of the total surface area of the activated carbon. Typically, the diameters of the micropores are less than 2 nm and the macadamia's activated carbon porous structure can be clearly seen in Fig. 12.

Figure 12(a) presents a cross-section view of the pore structure containing an array of pores. Figure 12(b) presents an enlarged image of a typical pore around 2μm in diameter. Inspection of the image also reveals an array of finer pores scattered over its inner surface. Studies by Poinern *et al.* have shown the importance of enhanced micropore development and the resultant improvement in gold adsorbing ability of an activated carbon [50]. Their investigation also found that higher activation temperatures ($\sim$900–1000°C) coupled with the resulting large surface areas had a significant impact on gold adsorption. For example, an activation temperature of 1000°C tended to produce an activated carbon capable of adsorbing 90% of the gold present in an aurocyanide solution within 8 h time period. Subsequent Energy Dispersive Spectroscopy and elemental mapping studies revealed the widespread presence of attached gold in the macadamia shell derived activated carbon. Their study also found that the macadamia shell derived activated carbon activated at 900°C and 1000°C was superior to commercially available coconut shell activated carbon [50]. However, their study also indicated that further research was needed to optimize the activation process to enhance the mechanical properties of the macadamia shell based activated carbon.

4.7 Concluding Remarks

Gold's allure has fired the imagination of humanity for thousands of years and continues to be a much sought-after metal. Today, the majority of the gold industry use a cyanide-based method for the extraction of gold from ores. Crucial to this globally used process is activated carbon with specific physical and chemical properties that have been engineered to enhance the adsorption of gold. However, today there is a need to find new sources of renewable and eco-friendly activated carbons. Relatively inexpensive, renewable and locally available agricultural waste materials have the potential to replace commercially imported activated carbons. Recent studies have shown promising results in producing activated carbons with large surface areas and high microporosities. The studies have shown that varying parameters such as activating agents, temperature and time can have a significant influence on properties such as pore size, porosity volume and surface area of the activated carbon. In particular, nutshells and stones from fruits produced by the food industry have a high carbon content and inherent strength that makes them an ideal precursor material for the production of activated carbons. Moreover, the constant and ever increasing supply of nutshells and stones offers an eco-friendly resource that can be converted into an economically viable product. However, before agricultural waste products such as nutshells and fruit stones can be totally converted into activated carbon products, there needs to be further detailed studies into optimizing material properties and developing cost-effective production techniques.

Acknowledgments

The authors would like to thank the members of the Murdoch Applied Nanotechnology Research group for their assistance and cooperation at various stages of the macadamia projects. The authors would also like to thank Horticulture Innovation Australia for a project Grant No. Al14003 that has partially funded this work.

References

[1] Puddephatt RJ (1978) The Chemistry of gold. Elsevier Scientific Publishing Company. The Netherlands.

[2] Schmidbaur H, Cronje S, Djordjevic B, Schuster O (2005) Understanding gold chemistry through relativity. *Chemical Physics*, **311**(1–2): 151–161.

[3] Winter M (1993) Gold: the essentials. Retrieved from http://www. webelements.com/gold/

[4] Gold Jewellery Alloys: (2000) Utilise Gold. Scientific, industrial and medical applications. World Gold Council. WWW.Utilisegold.com.

[5] Goodman P (2002) Current and future uses of gold in electronics. *Gold Bulletin*, **35**(1): 21–26.

[6] Corti CW, Holliday RJ, Thompson DT (2002) Developing new industrial applications for gold: gold nanotechnology. *Gold Bulletin*, **35**: 111–117.

[7] Niemeyer CM (2001) Nanoparticles, proteins, and nucleic acids: Biotechnology meets materials science. *Angewandte Chemie International Edition*, **40**: 4128–4158.

[8] Hirsch LR, *et al.* (2003) Nanoshell-mediated near-infrared thermal therapy of tumours under magnetic resonance guidance. *Proceedings of the National Academy of Sciences*, **100**: 13549–13554.

[9] Zheng Y, Sache L (2009) Gold nanoparticles enhance DNA damage induced by anti-cancer drugs and radiation. *Radiation Research*, **172**: 114–119.

[10] Seeger PA, Fowler WA, Clayton DD (1965) Nucleosynthesis of heavy elements by neutron capture. *The Astrophysical Journal Supplement Series*, **11**: 121.

[11] Willbold M, Elliott T, Moorbath S (2011) The tungsten isotopic composition of the Earth's mantle before the terminal bombardment. *Nature*, **477**(7363): 195–198.

[12] Hamburger M, Hereford A, Simmons W (Ed.) (2010) Volcanoes of the Eastern Sierra Nevada: Geology and Natural Heritage of the Long Valley Caldera.Bloomington, IN: Indiana University Department of Geological Sciences.

[13] Kettell B (1982) Gold, Oxford University Press, Melbourne, Australia.

[14] Griffith SV (1960) Alluvial prospecting and mining. 2nd Ed. Cox and Wyman Ltd, Great Britain.

[15] Gold Handbook (1983) Credit Suisse special publications, Vol. 66, Switzerland.

[16] Boyle RW (1987) Gold history and genesis of deposits. Van Nostrand Reinhold Company Inc. New York.

[17] Liddel K (1950) Handbook of non-ferrous metallurgy. McGraw-Hill Publishing Company, New York.

[18] Ferron CJ, Fleming CA (2003) Chloride as an alternative to cyanide for the extraction of gold – going full circle? SGS Minerals Services Technical Paper 01: 1–10.

[19] Fivaz CE (1988) How the MacArthur-Forrest cyanidation process ensured South Africa's gold future. *Journal of the Southern African Institute of Mining and Metallurgy* **88**(9): 309–318.

[20] Hodge SG, Ryan MG, Woodcock JT, Hamilton JK (1993) Australasian mining and metallurgy: The Sir Maurice Mawby Memorial Volume, Monograph 19, Parkville, 2, 1098.

[21] Marsden J, House I (1993) The Chemistry of gold extraction: Series in Metals and Associated Materials. Ellis Horwood Limited, New York.

[22] Kudryk V, Kellogg HH (1954) Mechanism and rate-controlling factors in the dissolution of gold in cyanide solution. *Journal of metals*, **6**: 541.

[23] Nicol MJ, Fleming CA, Paul RL, Stanley GG (1987) The extractive metallurgy of gold in South Africa. South Africa Institute of Mining Metallurgy, Johannesburg: pp. 831–905.

[24] Arbiter N, Han KN (Ed.) (1990) Gold: Advances in Precious Metal Recovery. Gordon and Breach Science Publishers.

[25] Dayha AS, King DJ (1983) The developments in carbon-in-pulp technology for gold recovery. *CIM Bulletin*, **76**(857): 55–61.

[26] Fleming CA (1998) The potential role of anion exchange resins in the gold industry. EPD Congress 1998. The Minerals, Metals and Materials Society, Warrendale, PA, USA: 95–117.

[27] Dicinoski GW (2000) Novel resins for the selective extraction of gold from copper rich ores. *South African Journal of Chemistry* **53**(1): 33–43.

[28] Van Deventer J, *et al.* (2000) Comparison of resin-in-solution and carbon–in–solution for the recovery of gold from clarified solutions. *Journal of the South African Institute of Mining and Metallurgy*, **100**(4): 221–227.

[29] Grosse C, Dicinoski GW, Shaw MJ, Haddad PR (2003) Leaching and recovery of gold using ammoniacal thiosulfate leach liquors (a review). *Hydrometallurgy*, **69**: 1–21.

[30] Schmitz PA, Duyvesteyn S, Johnson WP, Enloe L, McMullen J (2001) Ammoniacal thiosulfate and sodium cyanide leaching of preg-robbing Goldstrike ore carbonaceous matter. *Hydrometallurgy*, **60**: 25–40.

[31] Georgakilas V, Perman JA, Tucek J, Zboril, R (2015) Broad family of carbon nanollotropes: Classification, Chemistry, and applications of

fullerenes, carbon dots, nanotubes, graphene, nanodiamonds, and combined superstructures. *Chemical Reviews*, **115**(11): 4744–4822.

[32] Chung DDL (2004) Review: Electrical applications of carbon materials. *Journal of Materials Science*, **39**: 2645–2661.

[33] McDougall GJ (1991) The physical nature and manufacture of activated carbon. *Journal of the South African institute of mining and metallurgy*, **91**(4): 109–120.

[34] Biniak S, Szymanski G, Siedlewski J, Swiatkowski A (1997) The characterization of activated carbons with oxygen and nitrogen surface groups. *Carbon*, **35**(12): 1799–1810.

[35] Hayashi J, Yamamoto N, Horikawa T, Muroyama K, Gomes VG (2005) Preparation and characterisation of high-specific-surface-area activated carbons from K_2CO_3 treated waste polyurethane. *Journal of Colloid and Interface Science* **281**: 437– 443.

[36] Gadkaree KP (1997) Carbon honeycomb structures for adsorption applications: In Activated carbon compendium. Marsh H (Ed.), Elsevier Science Ltd, Oxford UK, 15–23.

[37] Marsh H (2001) Activated carbon compendium. Elsevier Science Ltd., Oxford UK.

[38] El–Geundi MS (1997) Adsorbents for industrial pollution control. *Adsorption Science and Technology* **15**: 777–787.

[39] Smisek M, Cerny S (1970) Active Carbon. Elsevier, New York.

[40] Zhang T, Walawender WP, Fan LT, Fan M, Daugaard D, Brown RC (2004) Preparation of activated carbon from forest and agricultural residues through CO_2 activation.*Chemical Engineering Journal*, **105**: 53–59.

[41] Armstrong CR (2004) Global demand for activated carbon increasing. *Public Works ProQuest Science Journals*, **135**(10): 22.

[42] Australian Bureau of Statistics (2000) Australian Bureau of Statistics — Import data on activated carbon. Canberra, ABS.

[43] Activated Carbon Market (Powdered, Granular) for Liquid Phase and Gas Phase Applications in Water Treatment, Food & Beverage Processing, Pharmaceutical & Medical, Automotive and Air purification — Global Industry Analysis, Size, Share, Growth, Trends and Forecast, 2013–2019.

[44] Ahmadpour A, Do DD (1997) The preparation of activated carbon from macadamia nutshell by chemical activation. *Carbon*, **35**: 1723–1732.

[45] Martınez ML, Torres MM, Guzman CA, Maestri DM (2006) Preparation and characteristics of activated carbon from olive stones and walnut shells.*Industrial Crops and Products*, **23**: 23–28.

[46] Tancredi N, Medero N, Möller F, Piriz J, Plada C, Cordero T (2004) Phenol adsorption onto powdered and granular activated carbon, prepared from eucalyptus wood. *Journal of Colloid and Interface Science* **279**: 357–363.

[47] Olivares-Marın M, Fernandez-Gonzalez C, Macıas-Garcıa A, Gomez-Serrano V (2006) Preparation of activated carbon from cherry stones by chemical activation with $ZnCl_2$. *Applied Surface Science*, **252**: 5967–5971.

[48] Yalcin M, Arol AI (2002) Gold cyanide adsorption characteristics of activated carbon of non-coconut shell origin. *Hydrometallurgy*, **63**: 201–206.

[49] Wartelle LH, Marshall WE (2001) Nutshells as granular activated carbons: physical, chemical and adsorptive properties. *Journal of Chemical Technology and Biotechnology* **76**: 451–455.

[50] Poinern GEJ, Senanayake G, Shah N, Thi-Le XN, Parkinson G, Fawcett D (2011) Adsorption of the aurocyanide, $Au(CN)_2^-$ complex on granular activated carbons derived from macadamia nut shells — A preliminary study. *Minerals Engineering*, **24**: 1694–1702.

[51] Tseng RL (2007) Physical and chemical properties and adsorption type of activated carbon prepared from plum kernels by NaOH activation. *Journal of Hazardous Materials*, **147**(3): 1020–1027.

[52] Hu Z, Srinivasan MP (2001) Mesoporous high-surface-ore activated carbon. *Microporous and Mesoporous Materials*, **43**: 267–275.

[53] Zhang Y, Zheng J, Qu X, Chen H (2007) Effect of granular activated carbon on degradation of methyl orange when applied in combination with high-voltage pulse discharge. *Journal of Colloid and Interface Science* **316**: 523–530.

[54] Zabaniotou A, Maday P, Oudenne PD, Jung CG, Delplancke MP, Fontana A (2004) Active carbon production from used tyre in two-stage procedure: industrial pyrolysis and bench scale activation with $H_2O–CO_2$ mixture. *Journal of Analytical and Applied Pyrolysis* **72**: 289–297.

[55] Teng H, Ho JA, Hsu YF (1997) Preparation of activated carbon from bituminous coals with CO_2 activation — influence of coal oxidation. In: Activated carbon compendium.Marsh, H. (Ed.), 2001, Elsevier Science Ltd, Oxford UK, 15–23.

[56] Lartey RB, Acquah F, Nketia KS (1999) Developing national capability for manufacture of activated carbons from agricultural wastes, Pub: Ghana Engineer.

[57] Khalili NR, Campbell M, Sandi G, Golas J (1999) Production of mirco- and mesopores activated carbon from paper mill sludge — Effect of zinc chloride activation. In: Activated carbon compendium.Marsh, H. (Ed.), 2001, Elsevier Science Ltd, Oxford UK, 3–13.

[58] Allwar A (2012) Characteristics of pore structures and surface chemistry of activated carbons by physisorption, Ftirand Boehm methods. *IOSR Journal of Applied Chemistry*, **2**(1): 9–15.

[59] Martin MJ, Artola A, Balaguer MD, Rigola M (2002) Towards waste minimisation in WWTP: activated carbon from biological sludge and its application in liquid phase adsorption. *J. Chem. Technol. Biot.***77**: 825–833.

[60] Allwar A, Ahmad MN, MohdAsri MN (2008) Textural characteristics of activated carbons prepared from oil palm shells activated with $ZnCl_2$ and pyrolysis under nitrogen and carbon dioxide. *Journal of Physical Science*, **19**(2): 93–104.

[61] Lastoskie C, Gubbins KE, Quirke N (1993) Pore size distribution analysis of microporous carbons: a density Functional theory approach. *Journal of Physical Chemistry* **97**: 4786–4796.

[62] Shen W, Li Z, Liu Y (2008) Surface chemical functional groups modification of porous carbon. *Chemical Engineering Journal*, **1**: 27–40.

[63] Puziy AM, Poddubnaya OI, Martınez–Alonso A, Suarez–Garcıa F, Tasco JMD (2005) Surface chemistry of phosphorus–containing carbons of ligno-cellulosic origin. *Carbon*,**43**: 2857–2868.

[64] Bandosz TJ, Ania CO (2006) Surface chemistry of activated carbons and its characterization: In Activated Carbon Surfaces in Environmental Remediation.Bandosz TJ (Ed.) Elsevier Ltd. New York, USA, pp. 159–229.

[65] Biniak S, Szymanski G, Siedlewski J, Swiatkowski A (1997) The characterization of activated carbons with oxygen and nitrogen surface groups. *Carbon*, **35**(12): 1799–1810.

[66] Ahmedna M, Marshall WE, Husseinyc AA, Raod RM, Goktepea I (2004) The use of nutshell carbons in drinking water filters for removal of trace metals. *Water Research*, **38**: 1062–1068.

[67] Chandra TC, Mirna MM, Sudaryanto Y, Ismadji S (2007) Adsorption of basic dye onto activated carbon prepared from durian shell: Studies of adsorption equilibrium and kinetics. *Chemical Engineering Journal*, **127**: 121–129.

[68] Johnson WD (1894) U.S. Patent 522, 260.

[69] Zadra JB, Engel AL, Heinen HJ (1952) RI 4843, U.S. Bureau of Mines.

[70] Hassler JW (1963) Activated Carbon. Chem. Pub. Comp. New York, United States of America.

[71] Bansal R.C., Donnet J.B., Stoeckli F. (1988) Active carbon, Marcel Dekker, Inc. New York.

[72] Yehaskel A (1978) Activated carbon manufacture and regeneration, Noyes Data Corporation, New Jersey, USA.

[73] Serrano VG, Cuerda-Correa EM, Fernandez-Gonzalez MC, Alexandre-Franco MF, Macias-Garcia A (2004) Preparation of activated carbons from chestnut wood by phosphoric acid-chemical activation, Study of microporosity and fractal dimension. *Material Letter*, **59**: 846–853.

[74] MacDonald JAF, Quinn DF (1996) Adsorbents for methane storage made by phosphoric acid activation of peach pits. *Carbon*, **34**(9): 1103–1108.

[75] Ismadji S, Sudaryanto Y, Hartono SB, Setiawan LEK, Ayucitra A (2005) Activated carbon from char obtained from vacuum pyrolysis of teak saw: pore structure development and characterisation. *Bioresource Technology*, **96**: 1364–1369.

[76] Guo Y, Lua AC (2000) Effect of heating temperature on the properties of chars and activated carbons prepared from oil palm stones. *Journal of Thermal Analysis and Calorimetry*, **60**: 417–425.

[77] Onal Y, Akmil-Basar C, Sarici-Ozdemir C, Erdogan S (2007) Textural development of sugar beet bagasse activated with $ZnCl_2$. *Journal of Hazardous Materials*, **142**(1–2): 138–143.

[78] Jogtoyen M, Derbyshire F (1998) Activated carbon from yellow poplar and white oak by H_3PO_4 activation. *Carbon*, **36**(7–8): 1085–1097.

[79] McDougall GJ, Hancock RD (1980) Activated carbons and gold — a literature Survey. *Minerals Science and Engineering*, **12**: 85–89.

[80] McDougall GJ, Hancock RD (1981) Gold complexes and activated carbon. *Gold Bulletin*, **14**: 138–153.

[81] Dubinin MM (1966) Porous structure and adsorption properties of active carbons. Chemistry and Physics of Carbon, Vol. 2, Marcel Dekker, New York.

[82] International Union of Pure and Applied Chemistry (1972) Manual of symbols and technology:Appendix 2, Part 1, Colloidal and surface chemistry, Pure and Applied chemistry, Vol. 31.

[83] Kaneko K, Ishii C (1992) Superhigh surface area determination of microporous solids, Colloids and surfaces **67**: 203–212.

[84] El–Sheikh AH, Newman AP, Al-Daffaee HK, Phull S, Cresswell N (2004) Characterisation of activated carbon prepared from a single cultivar of Jordanian olive stones by chemical and physicochemical techniques. *Journal of Analytical and Applied Pyrolysis*, **71**: 151–164.

[85] Toles CA, Marshall WE, Johns MM (1999) Surface functional groups on acid-activated nutshell carbons. *Carbon*, **37**: 1207–1214.

[86] Ladeira ACQ, Figueira MEM, Ciminelli VST (1993) Characterisation of activated carbons utilised in the gold industry: Physical and chemical properties and kinetic study. *Minerals Engineering*, **6**(6): 585–596.

[87] Muir DM (1982) Recovery of gold from cyanide solutions using activated carbon: A review, in Carbon in Pulp technology for the extraction of gold 1982, The Australian institute of mining and metallurgy, Australia.

[88] Fleming CA (1998) Thirty Years of turbulent change in the gold industry. *CIM Bulletin*, **91**: 55–67.

[89] Muir DM (1991) Principles and applications of carbon technology for gold recovery, Proceedings China Gold Conference, Tsingdao, China.

[90] Nicol MJ, Fleming CA, Cromberge G (1984) The adsorption of gold cyanide onto activated carbon. 1. The kinetics of adsorption from pulps. *Journal of the South African Institute of Mining and Metallurgy* **84**: 50–54.

[91] Nicol MJ, Fleming CA, Cromberge G (1984) The adsorption of gold cyanide onto activated carbon: Application of the kinetic model to multistage adsorption circuits. *Journal of the South African Institute of Mining and Metallurgy* **84**: 85–93.

[92] Dai X, Breuer PL, Jeffrey MI (2008) Micro-tomography based identification of the mechanisms of gold adsorption onto activated carbon and modelling. Young AC, Taylor PR, Anderson CG, Chi Y (Eds.), Hydrometallurgy 2008 — 6th International Symposium, SME, Littleton, 696–705.

[93] Demopoulos GP, Chen TC (2004) A case study of CIP tails slurry treatment comparison of cyanide recovery to cyanide destruction. *European Journal of Mineral Processing and Environmental Protection*, **4**(1): 1–9.

[94] Le Roux JD, Bryson AW, Young BD (1991) A comparison of several kinetic models for the adsorption of gold cyanide onto activated carbon. *Journal of the South African Institute of Mining and Metallurgy* **91**: 95–103.

[95] Jones WG, Klauber C, Linge HG (1988) The adsorption of gold cyanide onto activated carbon. Randol Gold Forum'88 Perth. Randol International, Golden, Co, 243–248.

[96] Jones WG, Klauber C, Linge HG (1989) Fundamental aspects of gold cyanide adsorption on activated carbon. World Gold'89. SME, Littleton, 278–281.

[97] Pleysier R, Dai X, Wingate CJ, Jeffrey MI (2008) Micro-tomography based identification of gold adsorption mechanisms, the measurement of activated carbon activity, and the effect of frothers on gold adsorption. *Minerals Engineering* **21**: 453–462.

[98] Gross J, Scott JW (1927) Precipitation of gold and silver from cyanide solution on Charcoal, U.S. Bureau of Mines, Technical Paper No. 378.

[99] Vegter NM (1992) The distribution of gold in activated carbon during adsorption from cyanide solutions. *Hydrometallurgy*, **30**: 229–242.

[100] Adams MD, McDougall GJ, Hancock RJ (1987) Models for the adsorption of aurocyanide onto activated carbon. Part II: Extraction of aurocyanide on pairs by polymeric adsorbents. *Hydrometallurgy*, **18**(2): 139–154.

[101] Adams MD (1990) The mechanism of adsorption of aurocyanide onto activated carbon — Relation between the effects of oxygen and ionic strength. *Hydrometallurgy*, 25 (2): 171–184.

[102] Davidson RJ (1974) The mechanism of gold adsorption in activated charcoal. *Journal of the South African Institute of Mining and Metallurgy*, **75**: 67–76.

[103] Adams MD, Friedl J, Wagner FE (1992) The mechanism of $Au(CN)^{4-}$ onto activated carbon. *Hydrometallurgy*, **31**: 265–275.

[104] McDougall GJ, Hancock RD, Nicol MJ, Wellington OL, Copperthwaite RG (1981) The mechanism of the adsorption of gold cyanide on activated carbon. *Journal of the South African Institute of Mining and Metallurgy* **80**: 344–356.

[105] Adams MD, Friedl J, Wagner FE (1995) The mechanism of adsorption of aurocyanide on to activated carbon: 2. Thermal stability of the adsorbed species. *Hydrometallurgy*, **37**(1): 33–45.

[106] Tsuchida N, Muir DM (1986) Potentiometric Studies on the adsorption of Au $(CN)_2^-$ and $Ag(CN)_2^-$ onto activated carbon. *Metall. Trans. B*, **17**: 523–528.

[107] Tsuchida N, Muir DM (1986) Studies on the role of oxygen in the adsorption of Au $(CN)_2^-$ and $Ag(CN)_2^-$ onto activated carbon. *Metall. Trans. B*, **17**: 529–533.

[108] Kongolo K, Bahr A, Friedl J, Wagner FE (1990) Au Mossbauer studies of the gold species on carbons from cyanide solutions. *Metallurgical and Materials Transactions B*, **21**: 239–249.

[109] Cashion JD, McGrath AC, Volz P, Hall JS (1988) Direct analysis of gold species on activated carbon by Mössbauer spectroscopy. *Transactions of The Institution of Mining and Metallurgy Section C*, **97**: C129–C133.

[110] Klauber C (1991) X-ray photoelectron spectroscopic study of the adsorption mechanism of aurocyanide onto activated carbon. *Langmuir*, **7**: 2153–2159.

[111] Ibrado AS, Fuerstenau DW (1995) Infrared and X-ray photoelectron spectroscopy studies on the adsorption of gold cyanide on activated carbon. *Minerals Engineering*, **8**: 441–458.

[112] Sibrell PL, Miller JD (1992) Significance of graphitic structural features in gold adsorption by carbon. *Minerals and Metallurgical Processing*, **9**: 189–195.

[113] Poinen G, Thurgate S (2003) Recovery of Gold from its Ores: An STM investigation of the adsorption of the aurocyanide ion onto Highly Orientated

Pyrolytic Graphite. In: Solid–Liquid Interfaces.Wandelt K, Thurgate S (Eds.), Springer–Verlag Berlin Heidelberg. Topics Applied Physics 85, 113–138.

[114] Poinern G, Thurgate SM, Kirton G, Ritchie IM (1998) Adsorption of dicyanoaurate (I) ions on highly oriented pyrolytic graphite. *Applied Surface Science* **134**: 73–77.

[115] Fleming CA (1982) Some aspects of the chemistry of carbon-in-pulp and resin-in-pulp processes. Carbon-in-pulp Technology for the Extraction of Gold. Aus.IMM, Melbourne, 415–440.

[116] Fleming CA, Nicol MJ, Nicol DJ (1980) The optimization of a carbon-in-pulp adsorption circuit based on the kinetics of extraction of aurocyanide by activated carbon. In: Presented at Mintek Meeting, Ion Exchange and Solvent Extraction in Mineral Processing. February, Randburg, Mintek.

[117] Fleming CA, Mezei A, Bourricaudy E, Canizares M, Ashbury M (2011) Factors influencing the rate of gold cyanide leaching and adsorption on activated carbon, and their impact on the design of CIL and CIP circuits. *Minerals Engineering*, **24**: 484–494.

[118] La Brooy SR, Bax AR (1985) Fouling of activated carbon by organic reagents, Proceedings of the 13th Australian Chemical Engineering Conference, Perth, Western Australia, 187–191.

[119] La Brooy SR, Bax AR, Muir DM, Hosking JW, Hughes HC, Parentich A (1986) Fouling of activated carbon by circuit organics. In: Fivaz CE, King RP(Eds.), Gold 100. J. S. Afr. Inst. of Min. Metall. Johannesburg, South Africa, 123–132.

[120] Salarirad MM, Behamfard A (2011) Fouling effect of different flotation and dewatering reagents on activated carbon and sorption kinetics of gold.*Hydrometallurgy*, **109**: 23–28.

[121] Le Roux JD, Bryson AW, Young BD (1991) A comparison of several kinetic models for the adsorption of gold cyanide onto activated carbon. *Journal of the South African Institute of Mining and Metallurgy* **91**: 95–103.

[122] American Society of Testing of Materials International — ASTM Designation D3802-79, April 1986, Standard test method for ball-pan hardness of activated carbons, USA. (www.astm.org).

[123] Avraamides J (1989) CIP carbons — selection, testing and plant operations. In: Bhappu, B. and Harden, R.J. (Eds.), Gold Forum on Technology and Practices – World Gold 89, Chapter 34, SME. Littleton, Colorado, 288–292.

[124] Cuhadaroglu D, Uygun OA (2008) Production and characterization of activated carbon from a bituminous coal by chemical activation. *African Journal of Biotechnology* **7** (20): 3703–3710.

[125] Campbell QP, Bunt JR, Kasaini H, Kruger DJ (2012)The preparation of activated carbon from South African coal. *Journal of the South African Institute of Mining and Metallurgy* **112**: 37–44.

[126] Khah AM, Ansari R (2009) Activated charcoal: preparation, characterization and applications: a review article. *International Journal of ChemTech Research* **1**(4): 859–864.

[127] Yahya MA, Al-Qodah Z, ZanariahNgah CW (2015) Agricultural bio-waste materials as potential sustainable precursors used for activated carbon production: A review. *Renewable and Sustainable Energy Reviews*, **46**: 218–235.

[128] Lozano-Castello D, Lillo-Rodenas MA, Cazorla-Amoros D, Linares-Solano A (2001) Preparation of activated carbons from Spanish anthracite. Activation by KOH. *Carbon*, **39**: 741–749.

[129] Donald J, Ohtsuka Y, Xu CC (2011) Effects of activation agents and intrinsic minerals on pore development in activated carbons derived from a Canadian peat. *Materials Letters* **65**: 744–747.

[130] Danish M, Hashim R, Ibrahim MNM, Rafatullah M, Ahmad T, Sulaiman O (2011) Characterization of Acacia mangium wood based activated carbons prepared in the presence of basic activating agents. *Bioresources*, **6**(3): 3019–3033.

[131] Patric JW (1995) Porosity in Carbons. Edward Arnold, London, U.K. 227–253.

[132] Ioannidou O, Zabaniotou A (2007) Agricultural residues as precursors for activated carbon production — a review. *Renewable & Sustainable Energy Reviews* **11**: 1966–2005.

[133] Madadi Yeganeh M, Kaghazchi T, Soleimani M (2006) Effect of Raw Materials on the Properties of Activated Carbons. *Journal of Chemical Engineering & Process Technology* **29**: 1247–1251.

[134] Cagnon B, Py X, Guillot A, Stoeckli F, Chambat G (2009) Contributions of hemi-cellulose, cellulose and lignin to the mass and the porous properties of chars and steam activated carbons from various lingo cellulosic precursors. *Bioresource Technology* **100**(1): 292–298.

[135] Daffalla SB, Mukhtar H, Shaharun MS (2012) Properties of activated carbon prepared from rice husk with chemical activation. *International Journal of Global Environmental Issues* **12**(2): 107–129.

[136] Ahmenda M, Marshall WE, Rao RM (2000) Production of granular activated carbons from selected agricultural by-products and evaluation of their physical, chemical and adsorption properties. *Bioresource Technology*, **71**: 113–123.

[137] Johns MM, Marshall WE, Toles CA (1998) Agricultural by-products as granular activated carbons for adsorbing dissolved metals and organics. *Journal of Chemical Engineering & Process Technology* **71**: 131–140.

[138] Soleimani M, Kaghazchi T (2005) The use of activated carbon prepared from agriculture waste in the separation of gold from industrial wastewaters. *Amirkabir Journal of Science & Research* **15**: 139–146.

[139] Ademiluyi FT, Amadi SA, Amakama Jacob N (2011) Adsorption and treatment of organic contaminants using activated carbon from waste Nigerian bamboo. *Journal of Applied Sciences and Environmental Management* **13**(3): 39–47.

[140] Hayashi J, Horikawa T, Muroyama K, Gomes VG (2002) Activated Carbon from Chickpea Husk by Chemical Activation with K_2CO_3: Preparation and Characterization. *Microporous Mesoporous Materials*, **55**: 63–68.

[141] Ekpete OA, Horsfall MJNR (2011) Preparation and characterization of activated carbon derived from Fluted Pumpkin Stem waste (Telfairiaoccidentalis Hook F). *Research Journal of Chemical and Environmental Sciences* **1**(3): 10–17.

[142] Wang J, Wu FA, Wang M, Qiu N, Liang Y, Fang SQ, *et al.* (2010) Preparation of activated carbon from a renewable agricultural residue of pruning mulberry shoot. *African Journal of Biotechnology* **9**(19): 2762–2767.

[143] Benaddi H, Bandosz TJ, Jagiello J, *et al.* (2000) Surface functionality and porosity of activated carbons obtained from chemical activation of wood. *Carbon*, **38**: 669–674.

[144] Gonzalez-Serrano ET, Cordero J, *et al.* (1997) Development of porosity upon chemical activation of Kraft Lignin with ZnCl2. *Industrial & Engineering Chemistry Research* **36**: 4832–4838.

[145] Arjmand S, Kaghazchi T, Mehdi Latifi S, Soleimani M (2006) Chemical production of activated carbon from nutshells and date stones. *Journal of Chemical Engineering & Process Technology* **29**: 986–991.

[146] Fouladi Tajar A, Kaghazchi T, Soleimani M (2009) Adsorption of cadmium from aqueous solutions on sulfurized activated carbon prepared from nutshells. *Journal of Hazardous Materials* **165**: 1159–1164.

[147] Aygün A, Yenisoy-Karaka S, Duman I (2003) Production of granular activated carbon from fruit stones and nutshells and evaluation of their physical, chemical and adsorption properties. *Microporous and Mesoporous Materials*, **66**(2–3): 189–195.

[148] Toles CA, Marshall WE, Johns MM (1998) Phosphoric acid activation of nutshells for metals and organic remediation: process optimization. *Journal of Chemical Technology and Biotechnology,* **72**: 255–263.

[149] Tongpoothorn W, Sriuttha M, Homchan P, Chanthai S, Ruangviriyachai C (2011) Preparation of activated carbon derived from jatropha curcas fruit shell by simple thermo-chemical activation and characterization of their physico-chemical properties. *Chemical Engineering Research and Design* **89**: 335–340.

[150] Toles CA, Marshall WE, Johns MM, Wartelle LH, McAloon A (2000) Acid-activated carbons from almond shells: physical, chemical and adsorptive properties and estimated cost of production. *Bioresource Technology* **71**: 87–92.

[151] Orkun Y, Karatepe N, Yavuz R (2012) Influence of temperature and impregnation ratio of H_3PO_4 on the production of activated carbon from hazel nutshell. *Acta Physica Polonica A* **121**: 277–280.

[152] Conesa JA, Sakurai M, Antal MJ (2000) Synthesis of a high-yield activated carbon by oxygen gasification of macadamia nutshell charcoal in hot, liquid water. *Carbon,* **38**(6): 839–848.

[153] Yang T, Lua AC (2003) Characteristics of activated carbons prepared from pistachio-nutshells by potassium hydroxide activation. *Microporous Mesoporous Materials,* **63**: 113–124.

[154] Ng C, Marshall WE, Rao RM, Bansode RR, Losso JN (2003) Activated carbon from pecan shell: process description and economic analysis. *Industrial Crops and Products,* **17**(3): 209–217.

[155] Kim JW, Sohn MH, Kim DS, Sohn SM, Kwon YS (2001) Production of granular activated carbon from waste walnut shell and its adsorption characteristics for Cu^{2+} ion. *Journal of Hazardous Materials* **85**: 301–315.

[156] Martinez ML, Torres MM, Guzman CA, Maestri DM (2006) Preparation and characteristics of activated carbon from olive stones and walnut shells. *Industrial Crops and Products,* **23**(1): 23–28.

[157] Genceli E, Apak E, Razvigorova M, Petrov N, Minkova V, Ekinci E (2002) Preparation, modification and characterization of pitches from apricot stones. *Fuel Process. Technol.* **75**: 97–107.

[158] Soleimani M, Kaghazchi T (2008) The investigation of the potential of activated hard shell of apricot stones as gold adsorbents. *Journal of Industrial and Engineering,* **14**: 28–37.

[159] Gergova K, Eser S (1996) Effects of activation method on the pore structure of activated carbons from apricot stones. *Carbon,* **34**: 879–888.

[160] Olivares-Marín M, Fernández-González C, Macías-García A, Gómez-Serrano V (2012) Preparation of activated carbon from cherry stones by physical activation in air. Influence of the chemical carbonisation with H₂SO₄. *Journal of Analytical and Applied Pyrolysis*, **94**: 131–137.

[161] El-Sheikh AH, Newman AP, Al-Daffaee HK, Phull S, Cresswell N (2004) Characterisation of activated carbon prepared from a single cultivar of Jordanian olive stones by chemical and physicochemical techniques. *Journal of Analytical and Applied Pyrolysis*, **71**: 151–164.

[162] Lafi WK (2001) Production of activated carbon from acorns and olive seeds. *Biomass and Bioenergy*, **20**(1): 57–62.

[163] Hannafi NE, Boumakhla MA, Berrama T, Bendjama Z (2008) Elimination of phenol by adsorption on activated carbon prepared from the peach cores: modelling and optimisation. *Desalination*, **223**: 264–268.

[164] Tahvildari K, Bigdeli T, Esfahani SN, Farshchi M (2009) Optimization of activated carbon preparation from peach stone. *The Journal of Pure and Applied Chemistry Research* **11**: 47–55.

[165] Courtney P (2003) http://www.abc.net.au/landline/stories/s962653.htm

[166] Ergon Energy, 2002. Environmental Report: http://www.ergon.com.au/_data/assests/pdf_file/0015/6405/2002–Enviromental–Report.pdf).

[167] Rural Industries Research and Development Corporation (2009) http://www.ridc.gov.au/programs/established–rural–industry/pollution/macadamia.cfm.

[168] Australian Bureau of Statistics (2015).

[169] Nguyen C, Do DD (1995) Preparation of carbon molecular sieves by macadamia nutshells. *Carbon*, **33**(12): 1717–1725.

[170] Wartelle LH, Marshall WE, Toles CA, Johns MM (2000) Comparison of nutshell granular activated carbons to commercial adsorbents for the purge-and-trap gas chromatography of volatile organic compounds. *Journal of Chromatography A*, **879**(2): 169–175.

Chapter 5
Development of Novel Biosorbents for Gold and Their Application for the Recovery of Gold from Spent Mobile Phones

Katsutoshi Inoue, Manju Gurung, Hidetaka Kawakita, Keisuke Ohto, Durga Parajuli, Bimala Pangeni, and Shafiq Alam

5.1 Introduction

As a result of rapid technological development in recent years, an increasing number of novel and more economical advanced electronics have been continuously replacing older ones, and consequently increasing mass consumption while simultaneously changing the lifecycle of electronics due to the ever changing societal aspirations. Among such electronics, the use of mobile phones has rapidly grown in recent years, and has consequently led to a pressing issue in that they are much more frequently

K. Inoue*, M. Gurung, B. Pangeni, H. Kawakita and K. Ohto
Department of Applied Chemistry, Saga University
Honjo-machi 1, Saga 840-8502, Japan
e-mail: *kanoko1921@gmail.com
inoueka@cc.saga-u.ac.jp

D. Parajuli
National Institute of Advanced Industrial Science and Technology
Tsukuba, Japan

S. Alam
Department of Chemical and Biological Engineering
College of Engineering, University of Saskatchewan
57 Campus Drive, Saskatoon, SK S7N 5A9, Canada

replaced rather than refurbished and now constitute the fastest growing component of waste electric and electronic equipments (WEEE) [1]. WEEE are typically non-homogeneous and thus very complicated with regard to separation and purification [2, 3]. It is interesting to note that modern electronics contain more than 60 elements [3]. Mobile phones are the same in composition with other electronic devices, and consist of plastics, ceramics, glass, and metals. Because waste printed circuit boards (PCBs), a key constituent of WEEE, contain a variety of critical metals, such waste represents an important secondary resource for numerous metals. Consequently, in order to prevent resource depletion, recycling of WEEE for the recovery of valuable metals is a popular subject of increasing importance. Also, in terms of economics, the recovery of precious metals from WEEE is attractive because the content of precious metals in waste PCBs is almost 10 times higher than those of concentrated ores [4–6]. In the last two decades, much attention has been devoted to the development of techniques for the recycling of WEEE. In this context, the most active research on the recovery of metals from waste PCBs and electronic scraps has been focused on hydrometallurgical technologies because they are much more suitable than pyrometallurgical technologies for the treatment of complicated materials such as WEEE. Furthermore, they are more environmentally benign, predictable, and easily controlled under mild working conditions than conventional pyrometallurgical techniques [7–9]. Hence, in recent years, hydrometallurgical routes have gained popularity and are increasingly being preferred over pyrometallurgical routes due to lower energy costs, clean working conditions, higher selectivity for targeted metals, moderate operating temperature, easy plant operation, low emission of toxic materials, etc [10].

The hydrometallurgical process consists of the leaching and the recovery of metal values from a leach liquor. Commercial scales of various recovery processes have been employed; these include precipitation, solvent extraction, ion exchange, adsorption, etc. Among these, precipitation and solvent extraction are suitable for concentrated leach liquors; that is, it is well known that solvent extraction is suitable for solutions containing the targeted solutes at concentrations higher than around 500 mg/L. On the other hand, the separation processes using solid separating materials (solid/liquid separation) such as ion exchange as well as chelating ion exchange and adsorption are suitable for dilute solutions. The conventional ion exchange technique using ion exchange resins and chelating resins produced from petroleum as well as adsorption technique using activated carbon suffer from various drawbacks such as low selectivity to the targeted metals over other metals, necessitating the use of large amounts of adsorbents, thus considerably increasing the cost. Furthermore, the treatments of spent ion exchange resins and adsorbents after extensive operation represent an additional disposal concern. That is, because repetitive adsorption followed by elution is accompanied by many times of swelling and contraction, ion exchange resins and adsorbents gradually deteriorate, generating cracks internally and externally, which seriously impedes the smooth flow operation using columns packed with such resins and adsorbents. Such deteriorated resins and adsorbents are disposed of by land-filling at special dumping sites or by incineration, both of which are not environmentally benign.

In recent years, much attention has been focused on the use of various industrial wastes, agricultural byproducts, and biological materials for hydrometallurgical purposes as alternative materials for ion exchange and adsorption to the conventional ion exchange resins and adsorbents. Among these, biopolymers have received great deal of attention because they represent renewable resources and are more environmentally benign, biodegradable, easily incinerated, and biocompatible than conventional materials such as plastics produced from petroleum. Consequently, numerous approaches have focused on the development of cheaper and more effective adsorbents containing natural polymers such as polysaccharides including cellulose and chitin, lignin, various kinds of polyphenols, etc [11–14]. These biopolymers represent an interesting and attractive alternative for the use as adsorbents after simple chemical treatments or modifications. For example, although chitin exhibits only poor adsorption behaviors for the majority of metal ions, chitosan produced from chitin by hydrolysis using concentrated sodium hydroxide solution exhibits interesting adsorption behaviors for metal ions [15]. Also, although cellulose is quite inert for metal ions as it is, it exhibits a strong and selective adsorption behavior for gold(III) ion after the simple chemical treatment using concentrated sulfuric acid as will be mentioned in detail later. These phenomena are attributable to the changes in their particular chemical structure and physico-chemical characteristics, giving rise to reactive functional groups such as amine and ether groups in polymer chains, which results in high reactivity and excellent selectivity towards metal ions.

One of the big problems related to the recovery of precious metals including gold by means of solvent extraction and adsorption/ion exchange is the difficulty related to their stripping or elution from the loaded solvents and adsorbents. For example, one of the traditional recovery processes of gold from alkaline cyanide solution is adsorption using activated carbon or strongly basic anion exchange resins. In these cases, because the desorption or elution of the loaded gold is nearly impossible, they are incinerated leaving ashes containing gold. On the other hand, although aqueous mixtures of thiourea in hydrochloric acid solution may be effective for the stripping or the desorption of gold, palladium and platinum extracted or adsorbed from acidic solutions, their post-treatments are difficult in terms of recovery to the final products. Consequently, also in their recovery from acidic solutions, incineration is the only method used in the practice. From such viewpoints, the adsorbents prepared from biomass and biomass waste are quite suitable for such purposes because they are cheap and can be easily incinerated at relatively low temperatures leaving only small quantities of ashes rich in precious metals. Comparatively, the incineration of plastic ion exchange resins and activated carbon suffers from various issues related to the generation of tar and/or coke which are tedious for the post-treatments in addition to the large quantity of energy required for complete incineration. Therefore, adsorbents prepared from biomass possess many advantages over conventional adsorbents.

From such viewpoints, we have prepared various bioadsorbents from a variety of feed materials including various biomass waste for the recovery of gold from acidic aqueous solutions. Some of these adsorbents will be introduced as typical examples in the following sections.

5.2 Novel Bioadsorbents for Gold Prepared by the Authors

5.2.1 *Adsorbents of lignophenol, lignocatechol, lignopyrogallol* [16, 17]

Lignophenol, lignocatechol, and lignopyrogallol are biodegradable plastics consisting of lignin as polymer matrices. Lignin is a very complicated

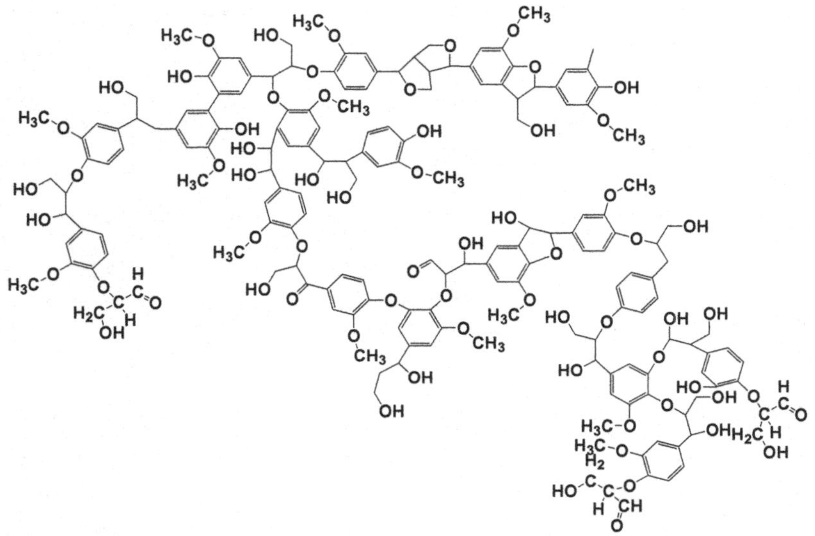

Scheme 1   Typical chemical structure of lignin

natural aromatic polymeric material as shown in Scheme 1. It is present in nearly all varieties of plants, such as trees and straws of rice and wheat in several 10% (20–40%) levels, depending on the species.

Lignin is easily extracted from nearly all types of plant-based biomass waste (i.e. saw dust) by interacting powdered feed materials with phenol or polyphenols such as catechol or pyrogallol in the form of lignophenol, lignocatechol, and lignopyrogallol, respectively, in the presence of concentrated sulfuric acid, during which polysaccharides such as cellulose are decomposed to be solubilized in the sulfuric acid solution and subsequently separated from lignin while the phenol or polyphenol molecules are immobilized onto polymer matrices of the lignin [18]. Scheme 2 illustrates the schematic flowsheet for the preparation of lignophenol. Thus, lignophenol, lignocatechol, and lignopyrogallol prepared in the current work were further cross-linked using paraformaldehyde in the presence of concentrated sulfuric acid to avoid dissolution in aqueous solutions. Scheme 3 shows the synthetic route of the cross-linked lignophenol as an example.

Although lignin itself exhibits only poor adsorption behaviors for metal ions, these properties can be greatly enhanced by immobilizing the

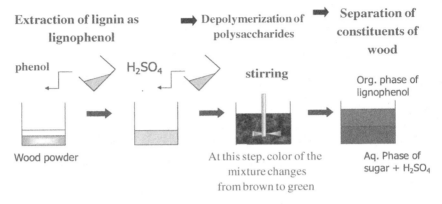

Scheme 2   Flowsheet of the preparation of lignophenol from wood powder (saw dust)

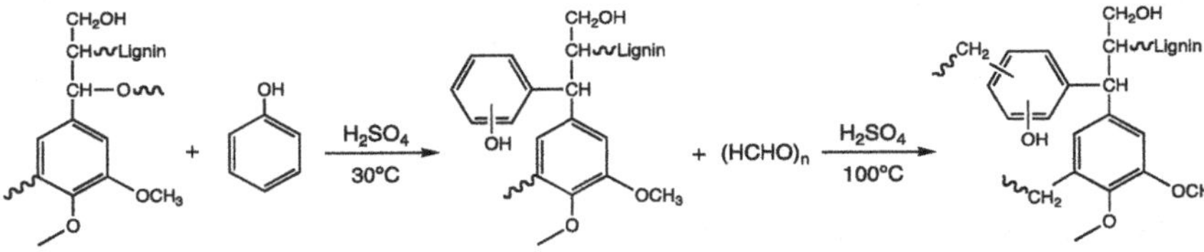

Scheme 3   Synthetic route of the cross-linked lignophenol

functional groups of phenol or polyphenols [16, 17, 19]. Such adsorption behaviors are attributable to phenolic hydroxyl groups of the immobilized functional groups of phenol and polyphenols.

Figure 1 shows the effect of hydrochloric acid concentration on the % adsorption of various metal ions on cross-linked lignophenol gel. As seen from this figure, only gold(III) is preferentially adsorbed on the gel without any significant co-adsorption of base metals such as copper(II) and iron(III) as well as other precious metals such as palladium(II) and platinum(IV), suggesting the extraordinary high selectivity of this gel to gold(III).

Figure 2 shows the scanning electron microscope (SEM) image of the filter cake of the cross-linked lignophenol gel recorded after the adsorption of gold (III). In this figure, very fine gold particles of significantly different shapes ranging from 10 μm to 50 μm are evidently observed to be scattered on the surface of the gel. The formation of elemental gold was also

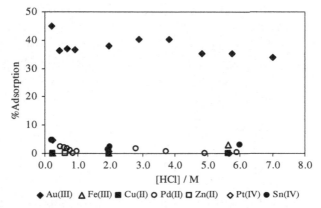

Fig. 1 Effect of hydrochloric acid concentration on the % adsorption of various metal ions on the cross-linked lignophenol gel [16]

Fig. 2 SEM image of the cross-linked lignophenol gel recorded after the adsorption of gold (III)

confirmed by means of X-ray diffraction (XRD), wherein strong peaks were observed at 38.18, 44.42, 64.68, and 77.5θ which belong to the crystal structure of elemental gold.

Many circumstances are considered responsible for the aforementioned phenomenon. Firstly, the gold(III) ion itself is a strong oxidizing agent with a standard reduction potential of $+1.40$ V vs. SHE. Another is the presence of a large number of phenolic hydroxyl groups along with carbonyl groups which tend to oxidize in aqueous medium in the presence of certain metal ions. Hence, a mutual electron transfer mechanism is believed to take place, thus establishing the redox system that provides electrons for the reduction of gold(III) ion as shown below.

$$\left(P\right)-\!\!\!\left\langle\!\!\!=\!\!\!\right\rangle\!\!-OH \ + H_2O \ = \ \left(P\right)-\!\!\!\left\langle\!\!\!=\!\!\!\right\rangle\!\!=O + 2H^+ + 2e^-,$$

$$Au^{3+} + 3e^- \ = \ Au^0.$$

Figure 3 shows the photograph of the solid/liquid mixture contained in a glass vessel after the adsorption of gold(III) on cross-linked lignophenol gel. Interestingly, characteristic gold colored aggregates of fine particles are visually observed to be floating on water surface. They are particles of metallic gold detached from the adsorption gel particles which have sunk down to the bottom of the glass vessel. It is inferred that the gold particles formed on the gel surface according to the above-mentioned mechanism are detached from the surface and aggregated at water surface as a consequence

Fig. 3 Photograph of the glass vessel after settling of the solid/liquid mixture after the adsorption of gold(III) on cross-linked lignophenol gel

of the hydrophobic nature of pure metal particles. By effectively taking advantage of such unique properties of gold for the particular system, it may be possible to recover gold(III) as pure metallic gold particles from various waste solutions containing dilute concentration of gold together with high concentration of other metals.

5.2.2 *Adsorbents of polysaccharides and biomass waste containing polysaccharides as major components* [20]

Numerous studies have been conducted on the adsorption of metals including gold using adsorption gels prepared from various polysaccharides such as chitosan [15]. These adsorption gels have been prepared by cross-linking the polymer materials of polysaccharides or through the immobilization of certain functional groups possessing higher affinities for metal ions. The purpose of the cross-linking is to prevent water solubility and improve the mechanical stability of the polymer materials of polysaccharides. Although some organic chemicals such as epichlorohydrin and glutaraldehyde have been employed as the cross-linking agents, we found that the use of a simple treatment involving boiling concentrated sulfuric acid is sufficient for the cross-linking conducted within our works [20]. Furthermore, although as mentioned earlier, the majority of polysaccharides themselves are inert for metal ions, we found that the treatment involving boiling concentrated sulfuric acid enables the drastic change in their adsorption characteristics as described in the following paragraph.

Figure 4 shows the % adsorption of some metal ions from varying concentrations of hydrochloric acid solution on the gel prepared from pure cellulose as a consequence of boiling concentrated sulfuric acid treatment. As seen from this figure, no metal ions are adsorbed except for gold(III), which is nearly quantitatively adsorbed over the whole concentration region tested, hence inferring the high selectivity of the gel for gold(III) over other metal ions tested, while, as well known, pure cellulose itself is inert for all metal ions including gold(III).

Figure 5 shows the adsorption isotherm of gold(III) at 303 K on the gel prepared from pure cellulose via the boiling concentrated sulfuric acid treatment. The amount of adsorption increases with increasing

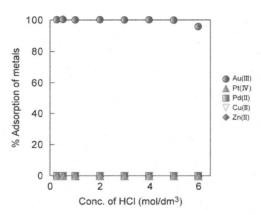

Fig. 4 Effect of hydrochloric acid concentration on the % adsorption of various metal ions on the gel prepared from pure cellulose by treating in boiling concentrated sulfuric acid

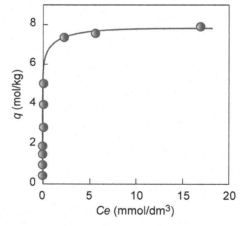

Fig. 5 Adsorption isotherm of Au(III) on the gel prepared from pure cellulose by treating in boiling concentrated sulfuric acid from 0.1 mol/dm³ HCl at 303 K

concentration of gold(III) at lower concentration whereas it tends to approach a constant value at high concentrations, suggesting monolayer adsorption according to the Langmuir model. The maximum adsorption capacity of this gel for gold(III) ion was evaluated from the constant value in the plateau region as 7.90 mol/kg (=1.55 kg/kg of dry gel), suggesting that higher quantity of gold than the weight of the adsorption gel itself was adsorbed on this gel.

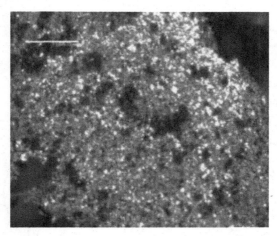

Fig. 6 Optical microscope photograph of the filter cake after the adsorption of gold(III) on the gel prepared from pure cellulose via the boiling concentrated sulfuric acid treatment

The exceptionally high selectivity and loading capacity of this gel for gold(III) may be attributed to the reduction of Au^{3+} to Au^0 during adsorption similar to the case of the adsorption on the cross-linked lignophenol gel mentioned earlier. Additionally, the formation of elemental gold was evidently confirmed through XRD analysis.

Figure 6 shows the microscope image of the filter cake after the adsorption of gold(III). In this figure, very fine aggregated particles of elemental gold with various shapes are distinctly visible. These results prove that gold particles formed by the reduction process have a certain tendency to be detached from the gel surface forming clear aggregates.

Similar phenomena were also observed for other polysaccharides such as dextran, pectic acid, and alginic acid. Such phenomena are believed to be attributed to some changes in the chemical structure of polysaccharides, making them suitable for coordinating gold(III), resulting in enhanced reactivity and excellent selectivity towards gold(III) as shown in Scheme 4 in the case of cellulose. The changes in the chemical structure were confirmed by means of infrared (IR) spectra, based on which the mechanism of adsorption of gold(III) was inferred as follows. The dissociated chain of polysaccharide in the presence of concentrated sulfuric acid undergoes the condensation cross-linking reaction with the release of water molecules.

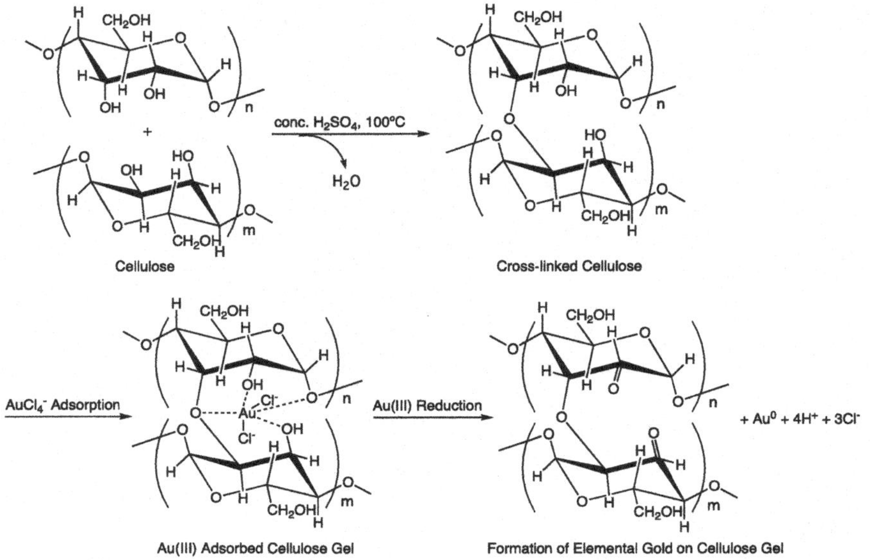

Scheme 4 Inferred example of the synthetic route of cross-linked cellulose and its adsorption–reduction mechanism [20]

Thus, the prepared gel is inferred to contain new ether bond linkages of C–O–C as observed in IR spectra. The oxygen atoms of these new C–O–C bonds play important roles as the coordination sites for gold(III) ion together with oxygen atoms of unreacted hydroxyl groups and those of –C–O–C bonds existing before the condensation reaction to form stable five-membered chelate rings. In acidic chloride media, the majority of gold(III) ion exists in the form of the tetra-chlorocomplex, $AuCl_4^{-}$. Furthermore, it is well known that gold(III) is easily reduced because it has high redox potential (1.0 V) in comparison to other metal ions like palladium(II) (0.68 V) and platinum(IV) (0.62 V).

Once in contact with the cross-linked polysaccharide gels, gold(III)–tetrachloro complexes are coordinated with the oxygen atoms of the sulfuric acid treated polysaccharides including those of ether bond. Furthermore, during this adsorption process, some hydroxyl groups are oxidized to carbonyl groups; consequently, such hydroxyl groups play an important role for the reduction of Au(III) to Au(0) as shown by the following reaction.

$$Au(III)_{(aq)} + R - OH \rightarrow Au^{(0)}_{(s)} + R = O + H^+.$$

Here, resulting carbonyl groups are easily protonated in acidic media and returned back into hydroxyl groups; that is, by repeating such reactions, the surface of the polysaccharide chemically modified by the concentrated sulfuric acid treatment is inferred to play the role of a catalyst for the reductive adsorption of gold(III) ion, which is considered to be the driving force for the formation of elemental gold particles, resulting in the extraordinary high selectivity and loading capacity.

Similar phenomena were observed in the cases of paper and cotton consisting of cellulose [21, 22], suggesting that this technology can be applied to various waste including waste papers, spent wears, and socks made of cotton or flax to produce the adsorption gels for gold recovery.

5.2.3 *Adsorbents of tannin compounds* [23]

Tannin compounds are polyphenol compounds possessing complicated chemical structures of many aromatic rings similar to lignin and are found in many varieties of plants including persimmon, tea leaves, grapes, chestnut pellicle, etc. Consequently, there are a variety of tannin compounds in nature depending on the particular plant species. Scheme 5 shows the chemical structure of tannin compounds contained in persimmon (persimmon tannin) [24].

Persimmon is a popular fruit in East Asian countries such as China, Korea, and Japan. There are two kinds of persimmon fruits, sweet persimmon and astringent persimmon, which differ by their content of water soluble persimmon tannin. Although the former can be eaten as they are similar to apples, the latter are rich in the water soluble persimmon tannin which generates a very astringent taste and are consequently unsuitable for direct consumption. However, the juice extracted from such astringent persimmon contains large quantities of persimmon tannin and has been traditionally employed in various applications such as natural paints, dyes, tanning agent for leather tanning, a coagulating agent for proteins, etc.

Sakaguchi and Nakajima found that some tannin compounds such as Chinese gallotannin (tannic acid) immobilized on cellulose and persimmon extract cross-linked using glutaraldehyde exhibit an extremely high affinity for some metal ions such as uranium [25]. They further found that the latter adsorbent is effective for the adsorption of gold(III) [26].

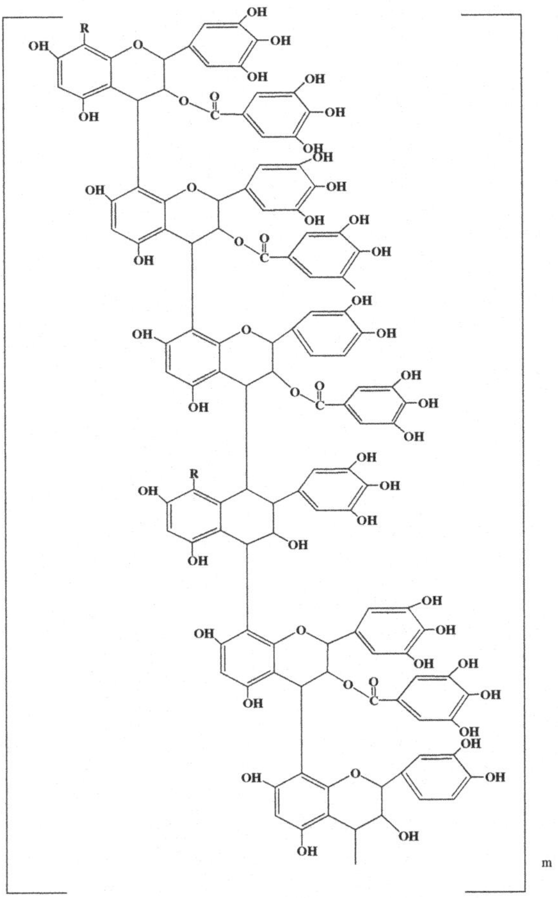

Scheme 5   Typical chemical structure of persimmon tannin [24]

As mentioned above, because persimmon tannin is partly soluble in water, it should be cross-linked to avoid the dissolution in aqueous solutions prior to be employed as an adsorbent. However, in our previous works, we found that the simple treatment in boiling concentrated sulfuric acid mentioned in the preceding section is also sufficient for such cross-linking. Scheme 6 illustrates the preparation route of such cross-linked persimmon tannin extract.

Here, the feed material was dried with persimmon tannin powder (abbreviated as PT powder, hereafter) extracted from astringent persimmon

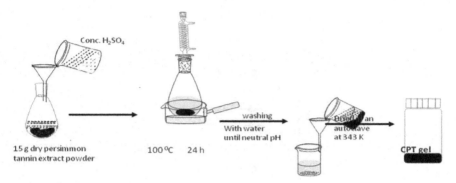

Scheme 6 Schematic preparation route of the adsorption gel (CPT gel) from persimmon tannin extract

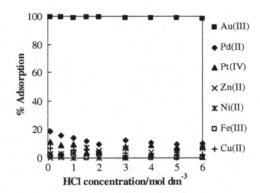

Fig. 7 % adsorption of various metal ions on the CPT gel from varying concentrations of hydrochloric acid [23]

and kindly donated by Persimmon-Kaki Technology Development, Co. Ltd., Jincheng, China. The resulting cross-linked persimmon tannin gel is abbreviated as CPT gel, hereafter.

Figure 7 illustrates the % adsorption of gold(III), platinum(IV) and palladium(II) as well as some base metals such as copper(II) from varying concentrations of hydrochloric acid on the CPT gel. Quantitative adsorption of gold(III) was achieved on the CPT gel over the whole concentration range of hydrochloric acid. Although about 10–20% adsorption was achieved for platinum(IV) and palladium(II), the adsorption of base metals was practically negligible under the present experimental conditions. The remarkably high selectivity of the CPT gel for gold(III) over base metals

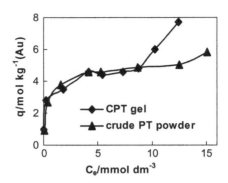

Fig. 8 Adsorption isotherms of gold(III) at 303 K on PT powder and CPT gel from 0.1 mol/dm³ hydrochloric acid solution [23]

as well as platinum(IV) and palladium(II) in a wide concentration range of hydrochloric acid is immensely useful with regard to the selective and quantitative recovery of gold(III) from acidic chloride media.

Figure 8 shows the adsorption isotherms of gold(III) at 303 K on the PT powder, the feed material, and CPT gel from the hydrochloric acid solution. From this figure, it is seen in both cases that amounts of gold adsorption were increased with increasing metal concentration of the test solution and tended to approach a constant value, after which they further increased, exhibiting a typical BET type adsorption isotherm based on the multilayer adsorption model. The gold uptake capacity of CPT gel was evaluated as high as 7.7 mol/kg (=1.52 kg-gold/kg-dry adsorbent), suggesting that greater quantities of gold(III) relative to the dry weight of these adsorbents were adsorbed, which are similar to the phenomenon observed in the case of polysaccharide gels cross-linked using boiling concentrated sulfuric acid. Also in this case, such extraordinary high adsorption capacities and selectivity towards gold(III) are attributable to the reductive adsorption phenomena of gold(III). The formation of elemental gold was confirmed by the observations by XRD and optical microscope.

In the case of the CPT gel, it is quite reasonable to infer that, in addition to the mechanism of the reductive adsorption of gold(III) by the polysaccharide contained in the persimmon extract together with persimmon tannin, various functional groups of such polyphenols existing in tannin molecules also play important roles as reducing agents, as pyrogallol is well known to function as a reducing agent.

Table 1   The maximum uptake capacities of gold(III) on various adsorbents

Adsorbent	Maximum uptake capacity [mol/kg]	Reference
Cross-linked lignophenol	1.9	[16, 17]
Cross-linked cellulose	7.6	[20]
Cross-linked paper	5.1	[21]
Cross-linked cotton	6.2	[22]
Persimmon extract powder (feed material of CPT)	5.9	[23]
CPT gel	7.7	[23]
Cross-linked persimmon waste after extracting persimmon tannin	5.0	[27]
Dimethylamine modified persimmon waste	5.6	[27]
Quaternary amine modified persimmon tannin	4.2	[28]
Cross-linked microalgal residue	3.3	[29]
Lysine modified cross-linked chitosan	0.35	[30]
Glycine modified cross-linked chitosan	0.86	[31]
Duolite GT-75	0.58	[32]
Dimethylamine modified waste paper	4.6	[33]

Table 1 shows the maximum uptake capacities of gold(III) on various adsorbents including those mentioned above for comparison. Among the adsorbents listed in this table, the CPT gel exhibits the highest uptake capacity.

By effectively using such noteworthy characteristics of various biomass, CPT gel in particular, we attempted to recover gold from various wastes including various e-wastes such as PCBs of spent mobile phones as a preliminary trial.

5.3   Recovery of Gold from PCBs of Spent Mobile Phones

5.3.1   *Pretreatment of spent mobile phones*

In the present study, as shown in Scheme 7, samples of waste PCBs from approximately 200 spent mobile phones were dismantled into some parts including PCBs by hand at the factory of SHIBATA INDUSTRY Co. Ltd., Omuta, Japan. Among them, the samples of PCBs were sent to the Shonan factory of TANAKA KIKINZOKU KOGYO K.K., Hiratsuka, Japan, where they were mechanically crushed using a two-axis crusher in order to effectively separate the rare earth magnets. Such crushed samples of

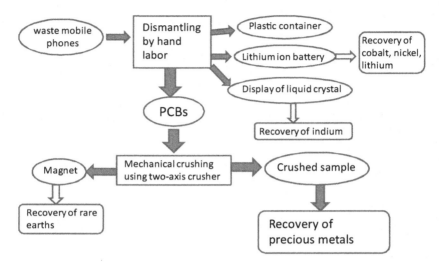

Scheme 7   Flowsheet of the dismantling of waste mobile phones into some parts

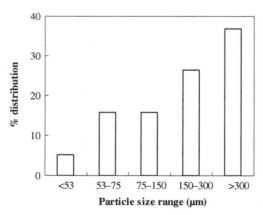

Fig. 9   Particle size distribution of the ash of the calcined sample of PCBs of spent mobile phones

PCBs were then calcined at 750°C for 6 h to remove epoxy resins, the main body of the PCBs on which various devices are attached, by incineration leaving metal values among the ash.

Figure 9 shows the particle size distribution of such ash of the calcined PCBs sample, which suggests that around 25% of the ash has a particle

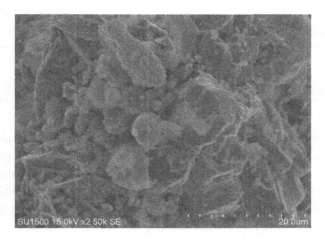

Fig. 10 SEM image of the ash of the calcined sample of PCBs of spent mobile phones

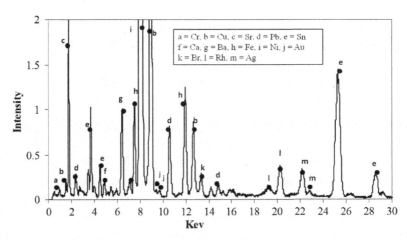

Fig. 11 EDX spectra of the ash of the calcined sample of PCBs of spent mobile phones

size smaller than 75 μm while the majority of the ash has that greater than 300 μm.

Figure 10 shows the SEM image of the ash, which reveals the extruded shape consisting of aggregated small particles with very irregular shapes.

Figure 11 shows the spectra obtained using energy-dispersive X-ray (EDX) spectroscopy of the ash contained in the calcined sample of PCBs

Table 2 Quantitative analysis of metals contained in the ash of the calcined sample of PCBs of spent mobile phones

Metals	Au	Pt	Pd	Ag	Cr	Fe	Co	Ni	Zn	Pb	Cu	Al
Content (mg/g)	1.63	0.01	0.39	0.52	0.11	10.5	0.20	19.1	1.69	13.2	349	0.69

of spent mobile phones, while Table 2 shows the result of the quantitative analysis of the same sample, which was measured using an inductively coupled plasma atomic emission spectrometer (ICP-AES) after totally dissolving the sample in boiling aqua regia. Both of these results indicate that copper is the main element existing in PCBs together with iron, lead, and nickel, contributing to the significant proportion of the metals in a relatively small portion of scrap. Nevertheless, it can be confirmed that PCBs of spent mobile phones are an outstanding secondary resource of these valuable metals.

The calcined sample was brought to the next step of leaching, in which metal values were extracted into aqueous solutions.

5.3.2 *Leaching of metal values from PCB of spent mobile phones*

There have been a variety of leaching technologies for precious metals. These include alkaline cyanide leaching for gold and silver and aqua regia leaching, which are typical conventional technologies commercially employed for a long time. However, these technologies suffer from various environmental problems such as their toxicity and the emission of NO_x gas. As new and emerging technologies, alternative to these conventional technologies, precious metals are leached using hydrochloric acid into which chlorine gas is blown, which is abbreviated as chlorine–HCl leaching, hereafter. Chlorine gas dissolved in hydrochloric acid is easily converted into hypochlorous acid and hydrochloric acid according to the following reaction.

$$Cl_2 + H_2O = HClO + HCl.$$

Here, hypochlorous acid functions as a strong oxidation agent, which dissolves metals in solid state into aqueous solutions, converting into metal ions. Such dissolved metal ions interact with hydrochloric acid or chloride ion forming stable metal-chloro complexes such as $AuCl_4^-$, $PdCl_4^{2-}$, and $PtCl_6^{2-}$ as shown below in the case of gold.

$$2Au_{(s)} + 3HClO + 6H^+ = 2Au^{3+} + 3HCl + 3H_2O,$$

$$Au^{3+} + 4Cl^- = AuCl_4^-.$$

On the other hand, since hypochlorous acid is very labile, it is easily decomposed into hydrochloric acid after functioning as an oxidation agent. Consequently, the leach liquor using the chlorine gas containing hydrochloric acid is practically a hydrochloric acid solution containing large quantities of various metal ions. This new technology has been already commercially employed by many precious metals plants all over the world.

In the present work, prior to the chlorine–HCl leaching, the calcined sample was leached using nitric acid in order to recover silver which is insoluble in hydrochloric acid in advance. The leach residue containing gold, palladium, platinum, and base metals was calcined again under the same condition and, then, leached using chlorine–HCl solution, the acid concentration of which was around 3 mol/dm^3, as shown in Scheme 8. The metal ion concentrations in this leach liquor were as follows: copper(II) 3360, nickel(II) 369, iron(III) 2908, zinc(II) 40, platinum(IV) 340, palladium(II) 10, gold(III) 150 (unit: mg/dm^3).

5.3.3 Recovery of gold(III) by means of selective adsorption using persimmon tannin gels

From the above-mentioned leach liquor, gold(III) was recovered at first by means of selective adsorption using CPT gel and, then, platinum(IV) and palladium(II) were recovered using the chemically modified persimmon tannin gel immobilized with functional groups of bisthiourea which exhibit high selectivity to these metal ions [34].

Figure 12 shows the effect of solid/liquid ratio (= dry weight of the adsorbent/volume of the leach liquor) on the adsorption of various metals contained in the leach liquor on the CPT gel. As seen from this figure, gold(III) is highly selectively and quantitatively adsorbed over other metals

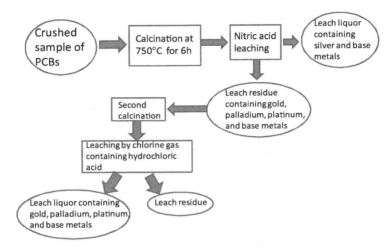

Scheme 8 Flowsheet of the leaching process of metal values from calcined sample of PCBs

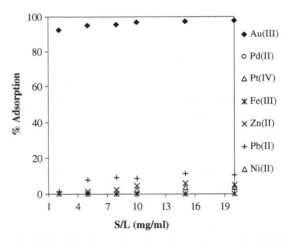

Fig. 12 Effect of solid/liquid ratio on the adsorptive recovery of gold(III) using CPT gel from the chloline–HCl leach liquor [34]

over a wide range of solid/liquid ratio as expected from the result of the fundamental adsorption test mentioned earlier.

On the basis of the above-mentioned result of the batchwise fundamental adsorption test, separation test of gold(III) from other metals was

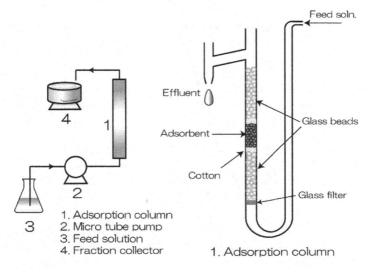

1. Adsorption column
2. Micro tube pump
3. Feed solution
4. Fraction collector

1. Adsorption column

Fig. 13 Schematic diagram of the packed column employed in the present work

carried out using a glass column packed with the CPT gel as shown in Fig. 13.

Figure 14 shows the breakthrough profiles of metal ions contained in the leach liquor from the column. As seen from this figure, the breakthrough of base metals took place immediately just after the start of flow, which is followed by those of palladium(II) and platinum(IV). The breakthrough of gold(III) began to take place after 200 BV. From this breakthrough profile of gold(III), the amount of gold(III) adsorbed on the bed was evaluated using 0.85 mol/kg-dry gel, which was less than that evaluated in the batchwise adsorption test.

After the finish of the breakthrough of gold(III), elution test was carried out using aqueous mixture of 0.5 mol/dm^3 hydrochloric acid and 0.5 mol/dm^3 thiourea as shown in Fig. 15. As seen from this figure, the elution of all metals except for gold(III) was not observed at all, i.e. only gold was eluted and concentrated as high as seven times higher than the feed solution. From the comparison with the adsorbed amount, it was found that 94.2% of the adsorbed gold(III) was eluted.

Although, as mentioned above, the gold(III) adsorbed on various bioadsorbents such as CPT gel can be easily eluted using aqueous mixture

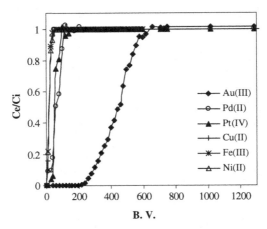

Fig. 14 Breakthrough profiles of gold(III) and other metals from the column packed with CPT gel for the chlorine–HCl leach liquor of PCBs of spent mobile phones. Ce and Ci stands for exit and inlet metal concentrations of the column. B.V. (bed volume) stand for the ratio, total volume of leach liquor which passed through the column/volume of the packed bed, corresponding to the time of the flow [34]

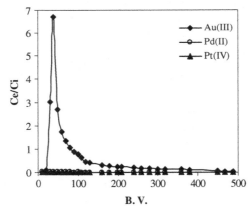

Fig. 15 Elution profile from the column packed with CPT gel loaded with gold(III) using eluting agent of the mixture of 0.5 mol/dm^3 hydrochloric acid and 0.5 mol/dm^3 thiourea [34]

of thiourea and hydrochloric acid, the easiest way for directly recovering metallic gold is the incineration of the gold loaded CPT gel making use of another characteristic of biomass, easy incineration at relatively low temperature (at around 500°C), which is one of their primary advantages

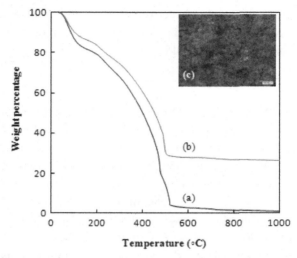

Fig. 16 TG analysis of cross-linked paper gel (a) before gold loading, (b) after gold loading, and (c) recovered gold after incineration [21]

over plastic ion exchange resins produced from petroleum and activated carbons.

For example, Fig. 16 shows the thermo-gravimetric (TG) curves of cross-linked paper gel before and after the gold loading, which illustrates the change in the percentage of the remaining weight of the gels as a function of the elevated temperature. In the case of the unloaded gel (curve (a)), the gel particles nearly vanished at high temperature. On the other hand, the TG curve of the gold loaded gel (curve (b)) suggests that the gel was only completely decomposed at temperature in excess of 500°C, leaving the weight (around 30% of the initial weight) of clear and pure gold aggregates as shown in (c), indicating how gold can be easily recovered with high purity from the gold loaded gel via simple incineration at high temperature. The weight loss at temperatures below 150°C is due to the evaporation of adsorbed H_2O molecules in both crude and gold loaded gel while the weight loss in the temperature range from 150°C to 500°C (gold loaded gel) and in that from 150°C to 520°C (non-loaded gel) is attributable to the decomposition of organic molecules by incineration reaction releasing H_2O and CO_2 molecules. Thus, it is expected that the gels of various biomass including CPT gel investigated in this work would show great promise as renewable materials capable of recovering elemental gold from industrial effluents such as those from gold plating and mining industries containing trace amount of the dissolved gold.

5.4 Conclusion

It was found that novel adsorption gels prepared from various biomass such as waste wood and straws, polysaccharides, including cellulose, as well as cellulose containing wastes such as spent cotton wears, socks and spent papers, and persimmon tannin exhibit very high selectivity to gold(III). Among these, those prepared from polysaccharides and persimmon extract by a simple treatment in boiling concentrated sulfuric acid exhibited very high selectivity and loading capacity, which is attributed to the adsorption mechanism for gold(III) coupled with its reduction reaction by many hydroxyl groups of polysaccharides and persimmon tannin, thus playing the role of a catalyst. By effectively using the remarkable properties of these adsorption gels, gold was able to be successfully recovered from actual leach liquor of PCBs of spent mobile phones. Although it was found that the adsorbed gold can be easily eluted using aqueous mixture of thiourea and hydrochloric acid, it was verified that the easiest way for the recovery as metallic gold is simple incineration at around 500°C, effectively using the combustible nature of biomass.

Acknowledgment

The authors are greatly indebted for the financial support for this study by the Grant-in-Aid for Scientific Research about establishing a Sound Material-Cycle Society by the Ministry of Environment of Japanese Government (K2131, K22075, K2354). We are also indebted to SHIBATA INDUSTRY Co. Ltd., Omuta, Japan, for the kind supply of the samples of PCBs of spent mobile phones as well as to Shonan factory of TANAKA KIKINZOKU KOGYO K.K., Hiratsuka, Japan, for the pretreatments and leaching of the PCBs.

References

[1] Robinson BH (2009) E-waste: An assessment of global production and environmental impacts. *Science of the Total Environment*, **408**(2): 183–191.

[2] Hagelueken C, Meskers C (2009) Technology challenges to recover precious and special metals from complex products, http://ewasteguide.info/files/Hagelueken_2009_R09.pdf

[3] Hagelueken C (2007) Metals recovery from e-scrap in a global environment. 6th Session of OEWG Basel Convention, http://archive.basel.int/industry/sideevent030907/unicor.pdf

[4] Li J, Lu H, Guo J, Xu Z, Zhou Y (2007) Recycle technology for recovering resources and products from waste printed circuit boards. *Environmental Science and Technology*, **41**: 1995.

[5] Guo J, Rao Q, Xu Z (2008) Application of glass-nonmetals of waste printed circuit boards to produce phenolic moulding compound. *Journal of Hazardous Materials*, **153**: 728.

[6] He W, Li G, Ma X, Wang H, Huang J, Xu M, Huang C (2006) WEEE recovery strategies and the WEEE treatment status in China. *Journal of Hazardous Materials*, **B136**: 502.

[7] Cui J, Zhang L (2008) Metallurgical recovery of metals from electronic waste: a review. *Journal of Hazardous Materials*, **158**: 228.

[8] Quinet P, Proost J, Lierde AV (2005) Recovery of precious metals from electronic scrap by hydrometallurgical processing routes. *Minerals and Metallurgical Processing*, **22**: 17.

[9] Kamberovic Z, Korac M, Ivsic D, Nikolic N, Ranitovic M (2009) Hydrometallurgical process for extraction of metals from electronic waste — part I. *Metallurgija-Journal of Metallurgy*, **15**: 231.

[10] Liew F.C. "Pyrometallurgy versus Hydrometallurgy", TES-AMM, Singapore, 2008 (http://www.tes-amm.com.au/downloads/TES-AMM_analysis_pyrometallurgy_vs_hydrometallurgy_April_2008.pdf)

[11] Thomas DA, Volesky B, Mucci A (2003) A review of the biochemistry of heavy metals biosorption by brown algae. *Water Research*, **37**: 4311.

[12] Marques PA, Pinheiro HM, Teixeira JA, Rosa MF (1999) Removal efficiency of Cu2+, Cd2+ and Pb2+ by waste brewery biomass: pH and cation association effects. *Desalination*, **124**: 137.

[13] Mack C, Wilhelmi B, Dancan JR, Burgess JE (2007) Biosorption of precious metals. *Biotechnology Advances*, **25**: 264.

[14] Volesky B, Holan ZR (1995) Biosorption of heavy metals. *Biotechnology Progress*, **11**: 235.

[15] Inoue K, Baba Y, Chitosan (2007) A Versatile Biopolymer for Separation, Purification, and Concentration of Metal Ions, In: Ion Exchange and Solvent Extraction, by A. Sengupta (ed.), Vol. 18, CRC Press, Boca Raton, pp. 339–374.

[16] Parajuli D, Inoue K, Kuriyama M, Funaoka M, Makino K (2005) Reductive adsorption of gold(III) by crosslinked lignophenol. *Chemistry Letters*, **34**: 34.

[17] Parajuli D, Adhikari CR, Kuriyama M, Kawakita H, Ohto K, Inoue K, Funaoka M (2006) Selective recovery of gold by novel lignin-based adsorption gel. *Industrial and Engineering Chemistry Research*, **45**: 8.

[18] Funaoka M (1998) A new type of phenolic lignin-based network polymer with the structure-variable function composed of 1,1-diarylpropane units. *Polymer International*, **47**: 277.

[19] Parajuli D, Inoue K, Ohto K, Oshima T, Murota A, Funaoka M, Makino K (2005) Adsorption of heavy metals on crosslinked lignocatechol: a modified lignin gel. *Reactive and Functional Polymers*, **62**: 129.

[20] Pangeni B, Paudyal H, Abe M, Inoue K, Kawakita H, Ohto K, Adhikari BB, Alam S (2012) Selective recovery of gold using some cross-linked polysaccharide gels. *Green Chemistry*, **14**: 1917.

[21] Pangeni B, Paudyal H, Inoue K, Kawakita H, Ohto K, Alam S (2012) An assessment of gold recovery processes using cross-linked paper gel. *Journal of Chemical and Engineering Data*, **57**: 796.

[22] Pangeni B, Paudyal H, Inoue K, Kawakita H, Ohto K, Alam S (2012) Selective recovery of gold(III) using cotton cellulose treated with concentrated sulfuric acid. *Cellulose*, **19**: 381.

[23] Gurung M, Adhikari BB, Kawakita H, Ohto K, Inoue K, Alam S (2011) Recovery of Au(III) by using low cost adsorbent prepared from persimmon tannin extract. *Chemical Engineering Journal*, **174**: 556.

[24] Matsuo T, Ito S (1978) The chemical structure of Kaki-tannin from immature fruit of the persimmon (*Diospyros kaki* L.). *Agricultural and Biological Chemistry*, **42**: 1637.

[25] Sakaguchi T, Nakajima A (1987) Recovery of uranium from seawater by immobilized tannin. *Seperation Science and Technology*, **22**: 1609.

[26] Nakajima A, Sakaguchi T (1993) Uptake and recovery of gold by immobilized persimmon tannin. *Journal of Chemical Technology and Biotechnology*, **57**: 321.

[27] Xiong Y, Adhikari CR, Kawakita H, Ohto K, Inoue K, Harada H (2009) Selective recovery of precious metals by persimmon waste chemically modified with dimethylamine. *Bioresource Technology*, **100**: 4083.

[28] Gurung M, Adhikari BB, Khunathai K, Kawakita H, Ohto K, Harada H, Inoue K (2011) Quaternary amine modified persimmon tannin gel: An efficient adsorbent for the recovery of precious metals from hydrochloric acid media. *Seperation Science and Technology*, **46**: 2250.

[29] Khunathai K, Xiong Y, Biswas BK, Adhikari BB, Kawakita H, Ohto K, Inoue K, Kato H, Kurata M, Atsumi K (2012) Selective recovery of gold by simultaneous adsorption – reduction using microalgal residues generated from biofuel conversion processes. *Journal of Chemical Technology and Biotechnology*, **87**: 393.

[30] Fujiwara K, Ramesh A, Maki T, Hasegawa H, Ueda K (2007) Adsorption of platinum(IV), palladium(II) and gold(III) from aqueous solutions onto L-lysine modified crosslinked chitosan resin. *Journal of Hazardous Materials*, **146**: 39.

[31] Ramesh A, Hasegawa H, Sugimoto W, Maki T, Ueda K (2008) Adsorption of gold(III), platinum(IV) and palladium(II) onto glycine modified crosslinked chitosan resin. *Bioresource Technology*, **99**: 3801.

[32] Idlesias M, Antico E, Salvado V (1999) Recovery of palladium(II) and gold(III) from diluted liquors using the resin Duolite GT-73. *Analytica Chimica Acta*, **381**: 61.

[33] Adhikari CR, Parajuli D, Kawakita H, Inoue K, Ohto K, Harada H (2008) Dimethylamine-modified waste paper for the recovery ofprecious metals. *Environmental Science and Technology*, **42**: 5486.

[34] Gurung M, Adhikari BB, Kawakita H, Ohto K, Inoue K, Alam S (2012) Selective recovery of precious metals from acidic leach liquor of circuit boards of spent mobile phones using chemically modified persimmon tannin gel. *Industrial and Engineering Chemistry Research*, **51**: 11901.

Chapter 6
Environmentally Friendly Processes for the Recovery of Gold from Waste Electrical and Electronic Equipment (WEEE): A Review

Isabella Lancellotti, Roberto Giovanardi, Elena Bursi, and Luisa Barbieri

6.1 Introduction

The recovery of gold from waste is more attracting with respect to other metals. Based on the data of World Gold Council, the request for gold has witnessed an increasing trend during the last decade [1]. This increase is not only due to the high request of the jewelry market but also due to the increasing demand for gold in industrial and medical applications.

Gold is commonly extracted from mines by mercury amalgamation and more refined techniques such as cyanidization [2], flotation, and smelting [3]. These are processes which have high impact on the environment, for example, mercury is a common by-product; besides the mercury amalgam-based processes, which are obsolete, Hg is often present in gold ores in huge amounts and can be released to the atmosphere during their processing,

I. Lancellotti, R. Giovanardi, E. Bursi and L. Barbieri
Department of Engineering "Enzo Ferrari",
University of Modena and Reggio Emilia, Modena, Italy
e-mail: isabella.lancellotti@unimore.it

for example, in refining steps of the extraction process if nothing is done to prevent it (this is a significant problem in countries where a proper legal framework is absent) [2, 4]. Anthropogenic mercury emissions are increasing since 1995. In 2010, it was estimated that 1960 tonnes of mercury were annually emitted into the air from all human activities [5].

In this context, in the last few years, many efforts were undertaken to design alternative environmentally acceptable procedures.

Hydrometallurgical processes have been exploited more often than pyrometallurgical processes. Pyrometallurgical processes includes many processes such as incineration, smelting in a plasma arc furnace or blast furnace, dressing, sintering, melting, and reactions in a gas phase at high temperatures. These pyrometallurgical processes are conventional methods to extract gold from ores and secondary raw materials. For example, in a pyrometallurgical process applied to secondary raw material, the crushed scraps containing gold are burned in a high temperature kiln and impurities are removed by means of many procedures, for example, they may be volatilized by a chemical reaction or by heat, or may be transformed into slags which rise to the surface of molten metal, or sludges which concentrate on the bottom [6–8].

The basic hydrometallurgy processing steps include extraction, concentration/purification, and recovery. The extraction consists of acid or alkaline leaching attacks of material containing gold. The eluates are then subjected to separation and purification steps such as precipitation of impurities, solvent extraction, adsorption, and ion exchange in order to obtain the metals of interest in an isolate and concentrate form. Subsequently, the solutions are treated by electrowining, cementation, chemical reduction, or crystallization for gold recovery [9].

For over 100 years, cyanide has been mainly exploited as a leaching reagent in gold mines and secondary sources because of its significant efficiency and relatively low cost, and around 18% of total production of cyanide is used in mining operation worldwide for extraction of gold. Considering that the increase in environmental disasters at various gold recovery plants around the world have caused severe contamination of natural water resources due to the use of cyanide as a leaching solution, more environmentally friendly leaching agents as substitutes are considered to overcome this serious concern.

Besides obtaining gold from ores, nowadays, it is particularly important to extract gold from secondary sources such as electronic and electric waste (WEEE). The treatment of WEEE is a current issue because the role of electric and electronic equipment in modern life is increasing. The rapidly changing technology renders gadgets and equipment obsolete thereby generating discarded electronic equipment, which are now the fastest growing waste stream in the world. Every year, 20–50 million tons of WEEEs are generated in the world, which could generate serious risks to human health and to the environment [10]. Due to the high amount of precious and non-precious metals utilized for electronic devices, the final disposal of this kind of wastes represents both an environmental and an economical concern.

The use of precious metals (PMs) in the electronic field is very important because PMs have unique properties, which give them a specific and irreplaceable role. Silver is the most currently used, over 5,000 tonnes silver per year are requested from the market, primarily for switches and contacts in electric equipments. Gold is the second PM in order of demand (around 250 tonnes/year and 8% of the total demand). It is present in almost every appliance in electronics where it is exploited for producing bonding wires in integrated circuits and/or it is used as a coating for contacts and connectors. Gold is used as solder in integrated circuits (in alloy with Sn) or for wide coatings where it guarantees protection against the corrosion. Palladium, platinum, rhodium, and iridium follow in order of consumption [11].

By considering the great amount of PMs contained in electronic devices, the recovery process of such waste seems economically sustainable and recovery of PMs, including gold from e-waste, is necessary in order to save the natural raw materials. For instance, the concentration of gold in gold ores is generally between 0.5 g and 15 g of metal per ton of mineral (0.5–15 ppm) [12], while in electronic circuit boards its concentration is 10 times higher (150 ppm in printed circuits board and higher than 10,000 ppm in Central Processing Units).

A detailed characterization of electronic scraps shows that gold is always present as thin coating over Cu, Fe, or Ag substrates, thus the removal of this supporting metal would provide a solid particulate which is composed of gold, plastic, and ceramic materials. This gold content would

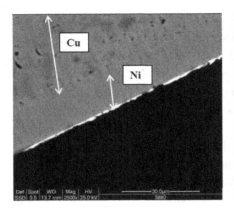

Fig. 1   SEM micrographs of electronic circuit boards scrap in section

then be easily separated by means of convenient physical techniques as those based on Eddy Current and Wet Eddy Current [13, 14].

SEM micrographs of waste printed circuit boards (PCB) used in this study (Fig. 1) show the presence of three layers of materials. In particular, the thin layer of gold can vary from 2.4 μm to 3.5 μm in old electronic circuit boards and 0.5–0.7 μm in the newest ones. The gold coating is supported by a copper layer (and often a Ni interlayer is present between them) applied on the plastic part of the board. Due to the irregular shape of the gold layer, the part of Ni and/or Cu can be in contact with leaching solutions used for gold extraction and can generate interference or, if the approach is to find leaching solutions to dissolve all the metals except Au, this irregular shape can be useful.

Nowadays, the recovery of metals from electronic wastes is generally performed by two strategies: by oxidative thermal treatment followed by metallurgical or chemical processes or by electrostatic separation of shredded boards [12, 15, 16]. For the last technique, the electrostatic separator separates conducting materials from non-conducting materials. The equipment has two electrodes of continuous current, one of induction and another of attraction. The significant difference in the electrical conductivity or specific electrical resistance between non-metals and metals gives a suitable condition for the implementation of a corona electrostatic separation in waste recycling. This separation method has been greatly utilized for the recovery of Cu or Al from electric wires and cables after grinding, and the recovery of Cu and PM from PCB scrap. Of course, this technique requires a preliminary size reduction of the wastes, using a method able to create scraps entirely (or mainly) composed of metals and scraps entirely (or mainly) composed of non-metals. Samples can be comminuted in a cutting mill, the types and sequence of mills are hammer mill firstly and then knife mill. At knife mill stage, different sizes of sieves are used to gradually decrease the size of particles. The smallest sieve is used to obtain particles size smaller than 1 mm [16, 17].

Every technique has some drawbacks as the first one consumes high amount of energy and solid such as non-combustible pollutant slag, and gaseous residues are produced, while the second procedure is not able to separate little amounts of metals from non-metallic supports, for this reason this technique is absolutely not suitable for the recovery of PMs contained in electronic devices.

More recently, *Magnetic and Electrical Separation*, magnetite nano-particles, synthesized through the co-precipitation method, were used to adsorb the positively charged complex of gold and thiourea obtained by dissolving gold in thiourea solution to investigate whether this new type of adsorbent was appropriate and efficient for the recovery of metallic gold. The novelties of this approach are firstly the high potential of such nanoparticles for gold recovery even at very low concentrations; second, their reproducibility after several recycling periods; and third, the very high efficiency [18].

Another very recent research study shows the use of ionic liquids for dissolution of gold. Ionic liquids are anhydrous salts that are liquid at low temperature [19].

They are powerful solvents and electrolytes with potential for high selectivity in both dissolution and recovery. Deep eutectic solvents (DESs) are a form of ionic liquid that are mixtures of salts such as choline chloride with hydrogen-bond donors such as urea. DESs are environmentally benign, yet chemically stable and, furthermore, the components are already produced in large quantities at comparable costs to conventional reagents. In contrast to aqueous liquids, where the solubility of metals is limited by the tendency of water to combine with metal ions and precipitate oxides and hydroxides, in water-free ionic liquids much higher metal concentrations can be achieved.

The eutectic mixtures have melting points that are significantly lower than their individual components, giving a room temperature ionic liquid. Their components are common, cheap chemicals, e.g. choline chloride (vitamin B4) is mainly used as an animal-feed additive, being already produced in large quantities.

Urea is a common nitrogen fertilizer that is non-flammable and completely biodegradable. Given that ionic liquids contain no water, it is not appropriate to refer to their application to ore and metal processing as hydrometallurgy. Instead, ionometallurgy [20] with DESs would seem to offer a new set of environmentally benign tools for metallurgists that could both augment existing processes and ultimately replace some.

6.2 Hydrometallurgical Process

Compared to pyrometallurgical processes, the hydrometallurgical method is more accurate, predictable, and it is easier to control it [21], but particular attention needs to be pointed to the kind of leaching solution used in the process. In Fig. 2, a flowchart of a general hydrometallurgical process is shown.

The first step of a hydrometallurgical process is *leaching*, an operation that exploits aqueous solutions to leach the metal of interest from a primary

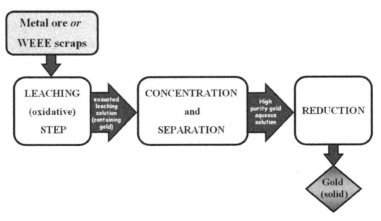

Fig. 2 Flowchart of a general hydrometallurgical process for the extraction and recovery of gold from ores or WEEE

source (ore) or a secondary source (waste) of the metal itself. The leaching solution features (that are specific for the metal to be extracted) can change in terms of pH, redox potential, presence of chelating agents and temperature: usually, these variables are set to optimize the leaching rate, as well as the selectivity of dissolution of the desired metal component into the aqueous phase.

After leaching, the aqueous solution obtained must normally undergo a *concentration* process to increase the relative content of the metal ions that are to be recovered; during this step, usually undesirable metal ions are removed. The *concentration* and (if needed) *separation* step can be accomplished through a wide selection of physical and chemical–physical techniques, such as precipitation, cementation, solvent extraction, and ion exchange [22].

Metal recovery is the final step in a hydrometallurgical process. The more important techniques for metal recovery processes are electrolysis, gaseous reduction, and precipitation. For example, copper is particularly extracted by hydrometallurgy procedures, because this metal is conveniently obtained by electrolysis. Cu^{2+} ions reduce at mild potentials, leaving behind other contaminating metals such as Fe^{2+} and Zn^{2+} [23].

Although there are several possibilities for the concentration, separation, and metal recovery steps of a hydrometallurgical process applied for the extraction of gold, the leaching is an extremely difficult step (considering the high nobility of gold and then its poor tendency to be oxidized and dissolved in solution as ion).

Significant interest has been shown to the use of alternative leaching agents to the conventional cyanidation process to extract gold from primary and secondary gold sources. This interest is mainly due to problems relative to the toxicity of cyanide and the lack of effectiveness of cyanidation process to leach, in an effective form, carbonaceous or complex ores.

Several alternatives to cyanide have been reported in Table 1. The table shows that thiosulfate solutions mixed with many other compounds are very attractive because thiosulfate is considered a non-toxic leaching agent and can leach gold in a faster way than cyanide. The leaching with thiosulfate could be evaluated as a good procedure for PM but with the problem of high cost raised in the purification step, it is difficult to apply this method at large scale.

Table 1 List of gold alternative leaching agents

Reagents	References
Ammoniacal thiosulfate	Lampinen *et al.* [24]
Thiourea–thiosulfate	Hongguang and David [25]
Ammoniacal/ammonium thiosulphate	Navarro *et al.* [26]
Thiourea	Jing-ying *et al.* [27]
1-Phenyl-2-thio-3-(2-hydroxyethyl) urea	Kwang *et al.* [28]
Acid thiourea	Örgül and Atalay [29]
Acidic thiourea with ferric iron	Behnamfard *et al.* [30]
Thiourea, chloride, (peroxomonosulfate ($HSO_5 -$ or iron(III) as oxidant) in 1-butyl-3-methylimidazolium hydrogen sulfate (bmimHSO$_4$), in 1-butyl-3-methylimidazolium hydrogen chloride (bmimCl) ionic liquid, in aqueous saturated K_2SO_4	Whitehead *et al.* [31]
Thiourea, bromide peroxomonosulfate ($HSO_5 -$ or iron(III) as oxidant in 1-butyl-3-methylimidazolium hydrogen sulfate (bmimHSO$_4$), in 1-butyl-3-methylimidazolium hydrogen chloride (bmimCl), in aqueous saturated K_2SO_4 as the solvent medium	Whitehead *et al.* [31]
Thiourea, iodide peroxomonosulfate ($HSO_5 -$ or iron(III) as oxidant in 1-butyl-3-methylimidazolium hydrogen sulfate (bmimHSO$_4$) in 1-butyl-3-methylimidazolium hydrogen chloride (bmimCl) in aqueous saturated K_2SO_4 as the solvent medium	Whitehead *et al.* [31]
Thiocyanate/thiourea	Zhang *et al.* [32]
Ethylene thiourea	McNulty [33]
Wet chlorination	Olteanu *et al.* [34]
Aqueous ozone–chloride	Viñals *et al.* [35]
Halides	Viñals *et al.* [35]
Bromine	Van Meersbergen *et al.* [36]
Acid ferric chloride	McNulty [33]
Iodide/supercritical water	Xiu *et al.* [37]
Iodide/iodine	Baghalha [38]
Bromine/bromide	McNulty [33]
Acid ferric chloride	McNulty [33]
High temperature chlorination	McNulty [33]
Aqua regia	Young and Derek [39]
	Lekka *et al.* [40]

(Continued)

Table 1   (*Continued*)

Reagents	References
Chloride–hypochlorite	Feng and van Deventer [41]
	Hasab *et al.* [42]
Ethaline mixture of choline chloride and ethylene glycol	Jenkin *et al.* [19]
1-alkyl-3-methyl-imidazolium ionic liquids	Whitehead *et al.* [43]
Thiocyanate	Li *et al.* [44]
Sodium sulfide	McNulty [33]
Alpha-hydroxynitriles	McNulty [33]
Malononitriles	McNulty [33]
Alkaline polysulfide	McNulty [33]
Alkaline glycine–peroxide	Oraby and Eksteen [45]
Natural organic acids	McNulty [33]
Alkali cyanoform	McNulty [33]

In the table, halides-based processes are also reported: these methods can be considered efficient as they present high leaching rates, but since the use of oxidant substances (usually the halogen of the halide itself) is needed, this would lead to high costs for the prevention of corrosion and the use of a bound system [22, 46–48].

Another group of leaching agents is constituted by thiourea and its admixture with other compounds: thiourea for the leaching of PM could be considered a proper method due to its high rate of leaching, low toxicity and eco-efficiency. Beside these advantages, this process presents the disadvantage that high quantity of reagent is consumed.

Figure 3 shows the main leaching agents for gold recovery present in literature from 2010 to 2015. Notwithstanding that cyanide is a toxic compound for humans and is harmful for the environment, it is still preferred and studied by several researchers and industries (about 25% of total research) because characteristics of effectiveness and low cost have contributed to make it a successful material worldwide for gold recovery. However, it should be underlined that research on cyanide is oriented especially on the extraction of gold from ores (about 96% of the studies) rather than from secondary sources, such as WEEE.

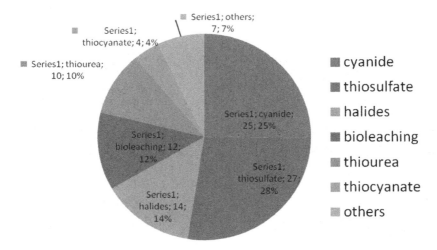

Fig. 3   Main leaching agents for gold recovery present in literature since 2010

Between non-cyanide lixiviants, thiosulfate has been the most studied, with a frequency in literature comparable to that of cyanide, with about 27% of total research. Among these, about 15% is directly applied to WEEE.

Thiourea is present in only 10% of the studies, but it should be noted that most of these (70%) concerns researches on WEEE, and in particular, PCB. Therefore, thiourea is currently considered as an interesting leaching agent for gold recovery from secondary sources.

Despite the potential advantages for the gold industry shown in the previous research, thiocyanate was the least employed between sulfur leachants in recent years (only 4% of works).

Bioleaching has also been very much investigated recently, with about 12% of total research. In this kind of technology, gold is bioleached via gold–cyanide complexation using cyanide produced from cultures of cyanogenic bacteria, such as *Bacillus megaterium, Chromobacteriumviolaceum, Pseudomonas aeruginosa, Pseudomonas fluroscens, Pseudomonas plecoglossicida* and fungi such as *Clitocybe sp., Polysporus sp.,* and *Marasmiusoreades*. Unlike the conventional hydrometallurgical processes for gold extraction with the direct use of cyanide ion, there are a number of works on bioleaching applied to WEEE [49–52].

Among halides (about 14% of total research), chlorination technologies have been the most applied (about 78% of works). The remaining works regard iodide and iodide–iodine leachants, while bromine has not been recently extensively studied for gold extraction.

Other researches regard technology based on the use of aqua regia, hydrogen peroxide, and amino acids such as glycine and ionic liquids.

6.3 Chloride-based Leachants

Although chloride-based leachants are not generally able to dissolve the metallic gold (except through the use of very strong, and dangerous, oxidizing agents, such as Cl_2 and Br_2 [53]), it is possible to detach the gold coatings of electric and electronic wastes (e.g. electronic circuit boards) using a chloride-based leachant capable of etching the metals that support the gold (copper, nickel, and iron).

This procedure [54] was developed considering the etching process used for the production of PCB, improving this method in order to operate with multi-metallic substrates and to be safer for the environment. Two chloride-based etching solutions may be used: a first one containing cupric chloride ($CuCl_2$) as oxidizing agent and a second one containing ferric chloride ($FeCl_3$); both the solutions operate in acidic conditions and in the presence of relatively high concentration of chloride ions, conditions guaranteed by the use of hydrochloric acid. The HCl prevents the precipitation of hydroxides, as $Fe(OH)_3$ and $Cu(OH)_2$; in addition, the chloride ions allow the formation of chlorinated complexes, ensuring that high concentration of metallic cations remains in the solution (formed during the etching).

The effectiveness of the two solutions is evaluated considering the rate with which they are able to bring into solution the more noble metal to be dissolved as a support to gold, i.e. the copper.

Considering the *copper dissolution rate* of the two leaching systems ($FeCl_3$-based and $CuCl_2$-based) it is possible to observe that: (i) in the $FeCl_3$-based process, the acid concentration has a minimal effect, while in the $CuCl_2$-based process, its contribution is important as much as the oxidant concentration; (ii) at low concentrations, the two systems have comparable etching rates, while at higher concentrations the $FeCl_3$-based process occurred up to three times faster. This result is explainable considering the standard reduction potentials of the two oxidizing agents: $E^0_{Fe^{3+}/Fe^{2+}} = +0.77$ V is greater than $E^0_{Cu^{2+}/Cu^+} = +0.16$ V, so the cupric ion is a weaker oxidant with respect to the ferric one.

In addition, to properly establish the chemical reactions that take place during the etching process, the influence of chloride anion as ligands must be considered. Cu^+ ion creates strong complexes with chloride ligands, so the prevalent forms of copper(I) in the studied process are probably $[CuCl_2]^-$ and $[CuCl_3]^{2-}$, while copper(II) creates weak complexes with chloride ligands (for example $[CuCl]^+$). This different stability of cupric and cuprous complexes enhance the electrochemical reduction of copper(II), raising its E^0. Also, the standard reduction potential of the iron (III)/iron(II) redox couple is affected by the chloride concentration, but in a more complex way: the reduction potential undergoes slight variations as long as the chloride concentration does not exceed the value of 1 M, after that the system increases its oxidizing power. Summarizing, the E^0 of the two redox couple (Fe^{3+}/Fe^{2+} and Cu^{2+}/Cu^+) is closer in the presence of chloride ions, and this justifies the comparable etching rates of the two systems; anyway, an increase in the concentration of chloride ions beyond the value of 1 M will raise the redox potential of the iron system, and justify the high effectiveness of the iron-based system (oxidation process up to three times faster).

Figure 4 shows the results obtained by applying this chemical leaching on CPU connection pins: in the image, a gold 'micro-tube' can be seen, obtained as a result of the complete dissolution of the internal steel support, a process that leaves the external gold coating perfectly undamaged. In particular, the leaching is obtained even when the oxidable metal is not directly exposed to the etching agent: probably, the wear of the external gold coating or some little scratches are enough to ensure that the reaction can proceed to the inside.

Besides the effectiveness of the leaching solution in terms of the etching rate, the recovery of the leachant is an aspect of great practical interest. In the process analyzed as the etching reaction proceeds the etching power of the solution fades, because the concentrations of cupric and/or ferric ions decrease (as they form the corresponding reduced form, Cu^+ or Fe^{2+}). This implies that, in order to use the etching solution for further leaching

Fig. 4   SEM micrographs of connection pins of a CPU after the chloride-based leaching process. SE detector

processes, it must be recovered to its initial "etching power". This should be performed using a readily available and cheap oxidizing reagent, avoiding the use of fresh strong oxidizers (such as Cl_2, H_2O_2) and the production of pollutant waste. The most appropriate oxidizer would be atmospheric oxygen (or better air).

Unfortunately, at the low pH assumed by these etching solutions, the oxidation of ferrous ion to ferric ion does not take place (even by bubbling of pure oxygen instead of air) being kinetically inhibited [55–57]; for this reason, in the case of iron, the recovery of the leachant is possible only through the use of strong oxidizing reagents, such as elemental chlorine, but this would make the process neither cheap nor environmental friendly.

Conversely, Cu^+ ions are easily (and completely) oxidized to Cu^{2+} by atmospheric oxygen, considering that the reaction is thermodynamically spontaneous and kinetically fast. In addition, during the leaching process of electronic components (such as PCB) using cupric-based solution, the metallic copper contained in the substrate (Cu^0) is oxidized to cuprous ions, so the exhausted leaching solution will contain a large amount of Cu^+ ions (they come both from oxidation of the substrate that from reduction of the oxidant). This allows one to consider the recovery of elemental copper by electrolysis before the restoring step. This electrolytic step is very useful

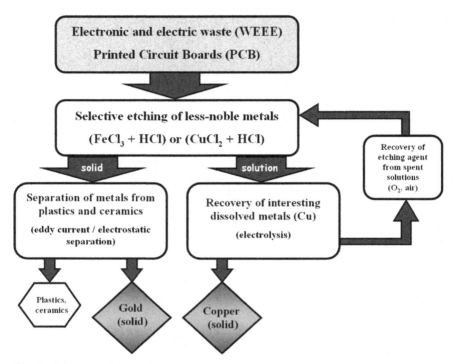

Fig. 5 Diagram of the *gold* and *copper* recovery process through PCB leaching using chloride-based etching solutions

as many of the metals which may have been dissolved during the etching process of electronic circuit boards (e.g. iron, tin, lead, aluminum, zinc) are stable in solution at the potential required for the reduction of Cu^+ to Cu^0, so only the reduction of the dissolved copper takes place.

The diagram of Fig. 5 schematically shows the whole process of recovery of gold and copper from electronic circuit boards (or other electrical/electronic waste) based on the method described. After the introduction of the electronic circuit boards directly in a suitable reactor containing the selected leaching solution (copper-based or iron-based), the etching process will start.

The reactor can be heated to accelerate the process; however, at room temperature, the complete detachment of the gold coating from the substrate (complete leaching of the copper underlayer), requires few minutes (the exact time depends on the type of oxidizing agent selected, its concentration and the pH of the etching solution). Once the detachment process is complete, it is possible to separate the different solid phases obtained (plastic, ceramic, and gold chips scraps) from the solution.

Using a simple filtration, it is possible to separate the solid phase from the liquid one, while using more sophisticated physical separation techniques exploiting different densities (floatation, centrifugation) or the ratio static charge/mass (electrostatic separation or eddy current techniques), it is possible to separate the gold solid phase from the plastic/ceramic one; the selection of the more suitable solid separation techniques strictly depends on the volume and the productivity of the plant.

Finally, the liquid phase undergoes a process of electrolysis to recover the copper dissolved from the electronic circuit boards, and then it is subjected to the reoxidation process (air insufflation) to restore the oxidizing agent.

6.4 Sulfur Leachants

6.4.1 *Thiosulfate*

Due to the low toxicity and the high sensitivity to gold cation, thiosulfate has been considered a potential non-cyanide leaching reagent for gold recovery, and in last decades it has been the subject of several studies. The gold dissolution by means of thiosulfate solution is generally slow, with relatively low yields and relatively high reagent consumption as compared to cyanide, making most thiosulfate systems not cost-effective for commercial application. Thiosulfate is able to form a stable anionic complex with Au^+ ion, as follows [22]:

$$4Au + O_2 + 8S_2O_3^{2-} + 2H_2O \rightarrow 4[Au(S_2O_3)_2]^{3-} + 4OH^-.$$

Cupric (Cu^{2+}) is added in thiosulfate leaching as catalyst in low concentration, and the leaching is performed in the presence of ammonia to stabilize copper, since Cu^{2+} can increase the decomposition of thiosulfate. Gold leaching in a thiosulfate solution containing cupric ion and ammonia is an electrochemical reaction according to the following equations [58]:

$$Au + 2S_2O_3^{2-} \rightarrow [Au(S_2O_3)_2]^{3-} + e^-,$$

$$[Cu(NH_3)_4]^{2+} + 3S_2O_3^{2-} + e^- \rightarrow 2[Cu(S_2O_3)_3]^{5-} + 4NH_3.$$

The concentration of Cu^{2+} ions in the thiosulfate solution is a significant factor affecting the reagent consumption, since it promotes the oxidative degradation of thiosulfate to tetrathionate, as follows:

$$2[Cu(NH_3)_4]^{2+} + 8S_2O_3^{2-} = 2[Cu(S_2O_3)_3]^{5-} + S_4O_6^{2-} + 8NH_3.$$

The oxidation rate of thiosulfate depends on ammonia concentration [59].

In particular, in the case of direct thiosulfate leaching of waste PCB, the copper dissolved from the PCB may adversely affect the leaching process through the decomposition of thiosulfate. Therefore, thiosulfate leaching should be applied after a preleaching stage in which copper was removed [58].

A study carried out on as-received sample of waste PCBs treated with an ammonium thiosulfate solution achieved only 16% gold recovered after 48 h of leaching. Nevertheless, a pretreatment of waste PCBs by milling accelerated the process, and 98% gold extraction after 48 h of leaching was achieved [60].

Experiments on gold leaching at varying temperature (20–50°C) were also conducted for a duration of 10 min using a solution containing 10 mM $CuSO_4$, 60 mM $Na_2S_2O_3$, 0.2 M NH_4Cl, and 0.26 M NH_3. From the results, it was stated that temperature is an important factor in gold leaching rate. In fact, while at 20°C the process yield was very low, almost total gold was dissolved in 2 min at above 40°C [61].

Other experiments on leaching of gold with sodium thiosulfate and ammonium thiosulfate showed that the best results were achieved when using a solution of 0.1 M of sodium thiosulfate, 0.2 M of ammonium hydroxide, and copper concentration between 0.015 M and 0.03 M, reaching about 15% of gold extraction [17].

6.4.2 *Thiourea*

In recent decades, thiourea $((NH_2)_2CS)$ has shown promising results in gold extraction from primary and secondary sources, and has been widely considered as a possible substitute of cyanide leachant since it presents several benefits [27]:

— fast kinetic reaction and high leaching rate;
— more environmental sustainability as compared to cyanide;

— more economical process as compared to thiosulfate;
— high selectivity to PM.

Studies on thiourea leaching suggest that the electron pairs of nitrogen and sulfur atoms have a better potential for a coordination bond with gold and silver, compared to cyanide [22].

Thiourea is able to form a stable cationic complex with gold [62]:

$$Au + 2SC(NH_2)_2 \rightarrow [Au(SC(NH_2)_2)_2]^+ + e^-.$$

However, a significant problem affecting the process is the reagent instability, since thiourea tends to decompose easily in alkaline medium. The reaction should occur in acidic solution accordingly. The kinetic of gold leaching with thiourea using different oxidants has shown that the most convenient is iron (Fe^{3+} ion). The following equation shows the gold dissolution in acidic solution of thiourea and ferric ion as oxidant [27, 63]:

$$Au + 2SC(NH_2)_2 + Fe^{3+} = [Au(SC(NH_2)_2)_2]^+ + Fe^{2+}.$$

However, ferric ion oxidizes thiourea easily in acidic solution forming formamidine disulfide, $(SCN_2H_3)_2$. This compound is an unstable product in acidic solution, and irreversibly decomposes to elemental sulfur and cyanamide.

$$(SCN_2H_3)_2 = SC(NH_2)_2 + NH_2CN + S.$$

Moreover, thiourea is consumed with the formation of a stable ferric sulfate product:

$$Fe^{3+} + SO_4^{2-} + SC(NH_2)_2 = [FeSO_4^*SC(NH_2)_2]^+.$$

Therefore, to avoid this problem, experimental conditions should be strictly controlled. This involves higher operating costs, as compared to cyanide [64].

Experiments of thiourea leaching were carried out on PCB materials from waste mobile phones changing different parameters, such as particle size, thiourea concentration, Fe^{3+} ion concentration, and temperature. The results obtained showed 90% Au recovery on particle size of 100 mesh in 24 g/l thiourea, 0.6% Fe^{3+} concentration at 25°C for 2 h leaching period.

It was observed that the gold dissolution rate reaches the maximum when thiourea content is 24 g/l. If thiourea concentration is too high, formamidine disulfide formation occurs. This compound is not stable at acid pH, and decomposes forming cyanamide and elemental sulfur. The problem is that, in addition to the reagent consumption, sulfur is able to form a thin film on gold surface, preventing its dissolution.

At the same time, Fe^{3+} is the oxidant, and if its concentration is too high, both gold and thiourea are oxidized (with formation of elemental sulfur and formamidine disulfide). When Fe^{3+} concentration is 0.6%, gold dissolution rate reaches the maximum.

Moreover, the higher the temperature, the more the thiourea decomposes. The best leaching temperature was thus found to be 25°C, since when it was lower, the reaction rate went slowly, negatively affecting gold dissolution [27].

Other leaching tests performed with thiourea and Fe^{3+} as oxidant on waste PCB showed that the Cu contained in PCB favors thiourea consumption preventing gold leaching. In fact, cupric ion oxidizes thiourea, but at the same time catalyzes its oxidation by Fe^{3+} and consequently its decomposition to glutinous elemental sulfur, which passivates gold surface. To avoid this problem, a preleaching of Cu was realized by means of a solution of H_2SO_4 and H_2O_2. It was observed that, more than an efficient Cu extraction, smallest particle size dimension should be ensured to reach a high gold leaching rate. The strong mixing dependence of gold leaching with thiourea was also observed, and so high speed mixing is required to have the best process efficiency [63, 64].

Figure 6 shows optical microscope photographs of a waste PCB pre (a) and post (b) a treatment with a solution of thiourea and ferric ion in the presence of copper. After the treatment, the gold surface is clearly passivated by a dark film, showing the degradation of thiourea solution.

In addition to the high costs and the high reagent consumption as compared to cyanide, the low chemical stability, acid thiourea leaching also involves the risk of carcinogenic effects, and so it must be treated with caution [22].

(a)   (b)

Fig. 6 Optical microscope photographs of a waste PCB pre- (a) and post- (b) treatment with a solution of thiourea and ferric ion in the presence of copper

6.4.3 *Thiocyanate*

Thiocyanate dissolves gold effectively in the presence of ferric ion, according to the following equations [44]:

$$3[Fe(SCN)_4]^- + Au = 3Fe^{2+} + [Au(SCN)_4]^- + 8SCN^-,$$

$$[Fe(SCN)_4]^- + Au = Fe^{2+} + [Au(SCN)_2]^- + 2SCN^-.$$

The acid system of thiocyanate and ferric ions has been considered as a potential substitute for cyanide in gold recovery, since it presents several benefits, such as high gold leaching rate and low reagent consumption [44].

Thiocyanate was able to recover above 95% of gold when KSCN concentration was 0.4 mol/L [22]. It was also observed that adding a small quantity of thiourea increases gold dissolution rate considerably. Therefore, it was possible to assume that thiourea catalyzes the process [44]. In a leaching study carried out on ores at room temperature, 57%, 66%, and 95% Au was extracted for leaching solutions containing thiourea only, ammonium thiocyanate only, and ammonium thiocyanate plus thiourea, respectively, under optimal conditions. It was found that the addition of thiocyanate reduces significantly thiourea consumption and therefore the process cost [32].

References

[1] World Gold Council (2001) Gold Demand Trends. 34 (February, 23).

[2] Korte F, Spiteller M, Coulston F (2000) The cyanide leaching gold recovery process is a non-sustainable technology with unacceptable impacts on ecosystems and humans: the disaster in Romania. *Ecotoxicology and Environmental Safety*, **46**(3): 241–245.

[3] Velásquez-López PC, Veiga MM, Mercury KH (2010) Balance in amalgamation in artisanal and small-scale gold mining: identifying strategies for reducing environmental pollution in Portovelo-Zaruma. *Ecuador Journal of Cleaner Production*, **18**(3): 226–232.

[4] UNEP: United Nations Environment Programme (2008) Mercury Use in Artisanal and Small Scale Gold Mining. Retrieved from http://www.unep.org (November 1, 2012).

[5] UNEP: United Nations Environment Programme (2013) Mercury Time to Act. Retrieved from http://www.unep.org (May 30, 2013).

[6] Hoffmann JE (1992) Recovering precious metals from electronic scrap. *Journal of the Minerals, Metals, and Materials Society*, **44**(7): 43–48.

[7] Lee JC, Song HT, Yoo JM (2007) Present status of the recycling of waste electrical and electronic equipment in Korea. *Resources, Conservation and Recycling*, **50**(4): 380–397.

[8] Sum EYL (1991) The recovery of metals from electronic scrap. *Journal of the Minerals, Metals, and Materials Society*, **43**(4): 53–61.

[9] Sadegh M, Safarzadeh MS, Bafghi D, Moradkhani (2007) A review on hydrometallurgical extraction and recovery of cadmium from various resources. *Minerals Engineering*, **20**(3): 211–220.

[10] Tuncuk A, Stazi V, Akcil A, Yazici EY, Deveci H (2012) Aqueous metal recovery techniques from e-scrap: hydrometallurgy in recycling. *Minerals Engineering*, **25**(1): 28–37.

[11] Bischoff A (2004) The use of precious metals in the electronics industry. In: Proceedings of the LBMA precious metals conference, Shanghai.

[12] Xiang D, Mou P, Wang J, Duan G, Zhang HC (2007) Printed circuit board recycling process and its environmental impact assessment. *International Journal of Advanced Manufacturing Technology*, **34**: 1030–1036.

[13] Cui J, Forssberg E (2003) Mechanical recycling of waste electric and electronic equipment: a review. *Journal of Hazardous Materials*, **99**: 243–263.

[14] Rem PC, Zhang S, Forssberg E, De Jong TPR (2000) Investigation of separability of particles smaller than 5 mm by Eddy-current separation

technology — part II: novel design concepts. *Magnetic and Electrical Separation*, **10**: 85–105.

[15] Scharnhorst W, Ludwig C, Wochele J, Jolliet O (2007) Heavy metal partitioning from electronic scrap during thermal end-of- Life treatment. *Science of the Total Environment*, **373**: 576–584.

[16] Veit HM, Diehl TR, Salami AP, Rodrigues JS, Bernardes AM, Tenorio JAS (2005) Utilization of magnetic and electrostatic separation in the recycling of printed circuit boards scrap. *Waste Management*, **25**: 67–74.

[17] Petter PMH, Veit HM, Bernardes AM (2014) Evaluation of gold and silver leaching from printed circuit board of cellphones. *Waste Management*, **34**: 475–482.

[18] Ranjbar R, Naderi M, Omidvar H, Amoabediny GH (2014) Gold recovery from copper anode slime by means of magnetite nanoparticles (MNPs). *Hydrometallurgy*, **143**: 54–59.

[19] Jenkin GRT, Al-Bassam AZM, Harris RC, Abbott AP, Smith DJ, Holwell DA, Chapman RJ, Stanley CJ (2015) The application of deep eutectic solvent ionic liquids for environmentally-friendly dissolution and recovery of precious metals. *Minerals Engineering*, In Press. 10.1016/j.mineng.2015.09.026

[20] Abbott AP, Harris RC, Holyoak F, Frisch G, Hartley J, Jenkin GRT (2015) Electrocatalytic recovery of elements from complex mixtures using deep eutectic solvents. *Green Chemistry*, **17**: 2172–2179.

[21] Andrews D, Raychaudhuri A, Frias C (2000) Environmentally sound technologies for recycling secondary lead. *Journal Power Sources*, **88**: 124–129.

[22] Syed S (2012) Recovery of gold from secondary sources — A review. *Hydrometallurgy*, **115–116**: 30–51.

[23] Habashi F (2009) Recent trends in extractive metallurgy. *Journal of Mining and Metallurgy, Section B: Metallurgy*, **45**: 1–13.

[24] Lampinen M, Laari A, Turunen I (2015) Ammoniacal thiosulfate leaching of pressure oxidized sulfide gold concentrate with low reagent consumption. *Hydrometallurgy*, **151**: 1–9.

[25] Hongguang Z, David BD (2002) The adsorption of gold and copper onto ion-exchange resins from ammoniacal thiosulfate solutions. *Hydrometallurgy*, **66**: 67–76.

[26] Navarro P, Vargas C, Villarroel A, Alguacil FJ (2002) On the use of ammoniacal/ammonium thiosulphate for gold extraction from a concentrate. *Hydrometallurgy*, **65**(1): 37–42.

[27] Jing-ying L, Xiu-li X, Wen-quan L (2012) Thiourea leaching gold and silver from the printed circuit boards of waste mobile phones. *Waste Management*, **32**: 1209–1212.

[28] Kwang SN, Byoung HJ, Jeon WA, Tae JH, Tam T, Myong JK (2008) Use of chloride–hypochlorite leachants to recover gold from tailing. *International Journal of Mineral Processing*, **86**: 131–140.

[29] Örgül S, Atalay Ü (2002) Reaction chemistry of gold leaching in thiourea solution for a Turkish gold ore. *Hydrometallurgy*, **67**: 71–77.

[30] Behnamfard A, Salarirad MM, Veglio F (2013) Process development for recovery of copper and precious metals from waste printed circuit boards with emphasize on palladium and gold leaching and precipitation. *Waste Management*, **33**(11): 2354–2363.

[31] Whitehead JA, Zhang J, McCluskey A, Lawrance GA (2009) Comparative leaching of a sulfidic gold ore in ionic liquid and aqueous acid with thiourea and halides using Fe(III) or HSO_5- oxidant. *Hydrometallurgy*, **98**(3–4): 276–280.

[32] Zhang J, Shen S, Cheng Y, Lan H, Hu X, Wang F (2014) Dual lixiviant leaching process for extraction and recovery of gold from ores at room temperature. *Hydrometallurgy*, **144–145**: 114–123.

[33] McNulty T (2001) Cyanide substitutes. *Mineralogical Magazine*, **184**(5): 256–261.

[34] Olteanu AF, Dobre T, Panturu E, Radu AD, Akcil A (2014) Experimental process analysis and mathematical modeling for selective gold leaching from slag through wet chlorination. *Hydrometallurgy*, **144–145**: 170–185.

[35] Viñals J, Juan E, Ruiz M, Ferrando E, Cruells M, Roca A, Casado J (2006) Leaching of gold and palladium with aqueous ozone in dilute chloride media. *Hydrometallurgy*, **81**: 142–151.

[36] Van Meersbergen MT, Lorenzen L, Van Deventer JSJ (1993) The electrochemical dissolution of gold in bromine medium. *Minerals Engineering*, **6**(8–10): 1067–1079.

[37] Xiu FR, Qi Y, Zhang FS (2015) Leaching of Au, Ag, and Pd from waste printed circuit boards of mobile phone by iodide lixiviant after supercritical water pre-treatment. *Waste Management*, **41**: 134–141.

[38] Baghalha M (2012) The leaching kinetics of an oxide gold ore with iodide/iodine solutions. *Hydrometallurgy*, **113–114**: 42–50.

[39] Young JP, Derek JF (2009) Recovery of high purity precious metals from printed circuit boards. *Journal of Hazardous Materials*, **164**: 1152–1158.

[40] Lekka M, Masavetas I, Benedetti AV, Moutsatsou A, Fedrizzi L (2015) Gold recovery from waste electrical and electronic equipment by electrodeposition: A feasibility study. *Hydrometallurgy*, **157**: 97–106.

[41] Feng D, Van Deventer JSJ (2006) Ammoniacal thiosulphate leaching of gold in the presence of pyrite. *Hydrometallurgy*, **82**: 126–132.

[42] Hasab MG, Raygan S, Rashchi F (2013) Chloride–hypochlorite leaching of gold from a mechanically activated refractory sulfide concentrate. *Hydrometallurgy*, **138**: 59–64.

[43] Whitehead JA, Zhang J, Pereira N, McCluskey A, Lawrance GA (2007) Application of 1-alkyl-3-methyl-imidazolium ionic liquids in the oxidative leaching of sulphidic copper, gold and silver ores. *Hydrometallurgy*, **88**(1–4): 109–120.

[44] Li J, Safarzadeh MS, Moats MS, Miller JD, LeVier KM, Dietrich M, Wan RY (2012) Thiocyanate hydrometallurgy for the recovery of gold. Part II: The leaching kinetics. *Hydrometallurgy*, **113–114**: 10–18.

[45] EA Oraby, JJ Eksteen (2015) The leaching of gold, silver and their alloys in alkaline glycine–peroxide solutions and their adsorption on carbon. *Hydrometallurgy*, **152**: 199–203.

[46] Yang H, Liu J, Yang J (2011a) Leaching copper from shredded particles of waste printed circuit boards. *Journal of Hazardous Materials*, **187**: 393–400.

[47] Yang X, Moats MS, Miller JD, Wang X, Shi X, Xu H (2011b) Thiourea–thiocyanate leaching system for gold. *Hydrometallurgy*, **106**: 58–63.

[48] Zhang Y, Liu S, Xie H, Zeng X, Li J (2012) Current status on leaching precious metals from waste printed circuit boards. *Procedia Environmental Sciences*, **16**: 560–568.

[49] Chi TD, Lee J, Pandey BD, Yoo K, Jeong J (2011) Bioleaching of gold and copper from waste mobile phone PCBs by using a cyanogenic bacterium. *Minerals Engineering*, **24**: 1219–1222.

[50] Natarajan G, Ting Y-P (2014) Pretreatment of e-waste and mutation of alkali-tolerant cyanogenic bacteria promote gold biorecovery. *Bioresource Technology*, **152**: 80–85.

[51] Natarajan G, Ting Y-P (2015a) Gold biorecovery from e-waste: An improved strategy through spent medium leaching with pH modification. *Chemosphere*, **136**: 232–238.

[52] Natarajan G, Tay SB, Yew WS, Ting Y-P (2015b) Engineered strains enhance gold biorecovery from electronic scrap. *Minerals Engineering*, **75**: 32–37.

[53] De Michelis I, Olivieri A, Ubaldini S, Ferella F, Beolchini F, Vegliò F (2013) Roasting and chlorine leaching of gold-bearing refractory concentrate: Experimental and process analysis. *International Journal of Mining Science and Technology*, **23**(5): 709–715.

[54] Barbieri L, Giovanardi R, Lancellotti I, Michelazzi M (2010) A new environmentally friendly process for the recovery of gold from electronic waste. *Environmental Chemistry Letters*, **8**: 171–178.

[55] España JS, Pamo EL, Pastor ES (2007) The oxidation of ferrous iron in acidic mine effluents from the Iberian Pyrite Belt (Odiel Basin, Huelva, Spain): field and laboratory rates. *Journal of Geochemical Exploration*, **92**: 120–132.

[56] Rao SR, Finch JA, Kuyucak N (1995) Ferrous-ferric oxidation in acidic mineral process effluents: comparison of methods. *Minerals Engineering*, **8**: 905–911.

[57] Stumm W, Lee GF (1961) Oxygenation of ferrous iron. *Industrial and Engineering Chemistry*, **53**: 143–146.

[58] Akcil A, Erust C, Gahan CS, Ozgun M, Sahin M, Tunkuc A (2015) Precious metal recovery from waste printed circuit boards using cyanide and non-cyanide lixiviant — A review. *Waste Management*, **45**: 258–271.

[59] Aylmore MG, Muir DM (2001) Thiosulfate leaching of gold — A review. *Minerals Engineering*, **14**(2): 135–174.

[60] Ficeriova J, Balazj P, Gock E (2011) Leaching of gold, silver and accompanying metals from circuit boards (PCBs) waste. *Acta Montanistica Slovaca*, **16**(2): 128–131.

[61] Ha VH, Lee J, Huynh TH, Jeong J, Pandey BD (2014) Optimizing the thiosulfate leaching of gold from printed circuit boards of discarded mobile phone. *Hydrometallurgy*, **149**: 118–126.

[62] Kulenov AS, Andreev AI, Pashkov GL, Kopanev AM, Belevantsev VI, Fedorov VA (1983) Formation complexes of gold(I) in aqueous solutions. *Zhurnal Neorganicheskoi Khimii*, **28**(9): 2418–2420.

[63] Birloaga I, De Michelis I, Ferella F, Buzatu M, Vegliò F (2013) Study on the influence of various factors in the hydrometallurgical processing of waste printed circuit boards for copper and gold recovery. *Waste Management*, **33**: 935–941.

[64] Birloaga I, Coman V, Kopacek B, Vegliò F (2014) An advanced study on the hydrometallurgical processing of waste computer printed circuit boards to extract their valuable content of metals. *Waste Management*, **34**: 2581–2586.

Chapter 7
Study on the Influence of Various Factors in the Hydrometallurgical Processing of Waste Electronic Materials for Gold Recovery

I. Birloaga and F. Veglià

7.1 Introduction

The devices of Information and Communications Technology (ICT) experienced, in the last two decades, the most growing demand in the market. This, correlated with the advances in the design of more innovative equipments and lowering of goods prices, had a main consequence as high reduction of lifespan for all the e-devices. Accordingly, a large amount of equipment is discharged and as no proper economic and environment-friendly technologies have been developed, most of all these wastes have been generally disposed of or incinerated considering the main core disintegration of their toxic content. With the increase in public concern over their environment and health issues, European Union Council has adopted two directives, namely, Directive on Waste Electrical and Electronic Equipments (WEEE) and Directive and Restriction of Hazardous

I. Birloaga* and F. Veglià
Department of Industrial and Information Engineering and Economics
University of L'Aquila, Italy
e-mail: *ionelapoenita.birloaga@graduate.univaq.it

Substances (RoHS). These two have been considered the base of legislation for the pre-production, production, and post-production of electronic wastes (e-wastes) and also as a starting point for the law application in China, Brazil, United States, and Canada [1]. It is estimated that about 5–8% of the total waste generated worldwide is represented by the e-wastes. According to Zoeteman *et al.* [2], the production of WEEEs of US, EU, Japan, and China is estimated to be of 5×10^6, 7×10^6, 3×10^6 and, respectively, 3×10^6 tons per year.

As in e-waste structure, a large amount of reusable components, not only valuable metals and materials but also toxic substances are found, thus the recycling of e-waste is considered as economic and eco-friendly. Considering this, the printed wiring board (PWB), a main embedded component within most of all the e-devices has attracted a large interest in recycling. This was mainly due to its large content of base (Cu, Fe, Ni, Sn, Zn, Pb, Al) and precious metals (PM) (Au, Ag, Pd). Since the concentration of Cu, Au, Ag, and Pd are considerably larger than their concentration into their natural minerals, these elements are considered the main economical drivers in the treatment of this valuable e-waste [3]. Beside this, other compositional materials of waste PWBs like polymeric and ceramic materials also present an interest in recycling as these may be used as a source of energy by their incineration and, respectively, after a total decontamination of metals, polymers, and other substances, as a raw material in the manufacturing of new products [4]. The printed circuit boards (PCBs) manufacturing has been known for its tremendous increase in the last 20 years. According to Prismark [5], the production trend since 1980–2012 has grown from 10,000 million to 60,000 million US$ and for the end of 2016 is estimated an increase of about 80,000 million US$. Single-layered to multi-layered or single-sided to double-sided with conducting layers of thin copper foil and insulating dielectric composite fibers represent the frame of PCBs. The flame retardant (FR) that represent the flammability safety of the epoxy resin composite fiber is generally of FR-4 for computers and communication equipment and FR-2 for home electronics and television. From the total of e-waste amount generated each year, about 3% are represented by the waste PCBs (WPCBs). For the processing of PWB, various technologies have been studied and some of them are even implemented at industrial level. Most of all these technologies are based on the recovery of the PM, particularly for gold that presents more than 75% of the economical value in the treatment of WPCBs [6].

In countries like China, India, Nigeria, and Ghana, where large amount of WEEEs are imported from different parts of the globe, the method of incineration is generally applied. This is mainly used for the recovery of energy but as is declared by Abreu and Negrão [7], "In order to produce energy, municipal solid waste cannot be of any type, it should be mostly plastics, dry papers, and petrochemicals". Therefore, these wastes as large amount of recyclable materials are lost by performing this operation and due to the Directives implemented at European level (these being transposed afterwards also at global level under various laws) that promotes the recycling of all materials, this technology cannot be anymore considered as a suitable methodology for processing of WEEE.

Presently, large companies like Umicore/Hoboken, Boliden, Noranda, and others have developed complexes and fluxes consisting of physical, pyrometallurgical, and hydrometallurgical procedures for the treatment of PWBs along with other industrial wastes.

In the scientific papers published by the researchers of Umicore, the process is based mainly on the Precious Metala Operations (PMOs) recovery procedure. Their process consists in processing of a mixture of various scraps (e-wastes, catalysts, tankhouse slimes, etc.), having in their composition a considerable content of PM and not only in a IsaSmelt furnance where these are smelted under air injection and with coke as reducing agent for base metals. In this way, the separation of PM in copper bullion by the base ones is achieved. The resulted bullion is subjected afterwards to leaching and electrowinning procedure for Cu recovery and the remained slag composed of PM is further refined by using cupellation procedure. However, all these technologies require large investments, i.e. Umicore has completed a sum of €200 million till 2006 to pass from the concentrates treatment to the processing of recyclable materials and some other industrial materials [8,9].

Similar procedures are also applied by the other above-mentioned large operators and considering high value of costs and investment, it is generally almost impossible to be undertaken for the start-up of such a complex process by any small and medium enterprise (SME).

Another process of e-waste processing, based mainly on the thermal treatment, is the pyrolysis procedure. This procedure was investigated and applied for both metal enrichment into the treated materials and energy savings considerate. It consists in use of special designed reactors (autoclaves) in which the already mechanically processed material (disassembled and milled) is treated at various temperatures. Application of this procedure produces both irreversible phase transformation and changes of chemical composition are simultaneous produced. Accordingly, the treatment of WPCB with this procedure leads to achieving of three products, namely, char (solid material), tar (oil), and gas (depending on the temperature level and used type of salt). The process takes place generally under an inert or oxidative atmosphere and in the presence of a molten salt [10–16]. However, this process leads to the separation of the organic materials and realization of a metal concentrate that further must be treated for the selective recovery of metals. Consequently, this procedure is generally considered as a suitable option in the pretreatment step of WPCBs.

Other process that may also be considered a suitable pretreatment procedure is the one of supercritical fluids technology. Many authors have involved the already mechanically shredded PCBs for the processing in supercritical water or a mixture of water with a chemical that is capable to do the separation of the epoxy resin by the PCB metallic part. The supercritical treatment with supercritical carbon dioxide was also investigated in the works published in the scientific literature.

Another procedure which was considered very efficient in the treatment of ores is the bio-hydrometallurgical procedure, in particular, bio-leaching process. This consists in using of various bacteria that is capable of dissolving selectively the elements of interests. In the treatment of e-wastes for gold extraction, the following microorganisms have been used: cyanogeum chromobacterium violaceum [17–22]; pseudomonas chlororaphis [23, 24]. Generally, this process is considered an eco-friendly procedure. However, the main inconvenience of this process is the very large time intervals required for metal extraction.

High interest of research has been geared towards hydrometallurgical procedures as these present the advantage of lower energy requirements and more easily handling operations. Below, a literature overview of the current aqueous technologies of gold from WPCBs is presented.

7.2 Gold: Generalities

Gold with the Latin symbol *aurum* (Au) (shinning down) is a dense metal of a bright yellow color which presents very good electrical and thermal properties and a melting point of 1064.18°C; this PM is known to the ancients (2600 B.C.). It has the atomic number of 79 with one stable isotope (^{197}Au) and an atomic mass of 196.96655. Au has a lattice structure of face-centred cubic (FCC) and constant of 4.040 (Å). Gold is extremely rare, being mostly founded in rocks in both native and compound forms (mostly as telluride — $AuTe_2$). The estimated amounts of gold recovered amounts and primary reserves at worldwide level since 2007–2013 are shown in Figs. 1 and 2. It has an average spreading into earth's crust of about 0.005 parts per million.

As can be seen within these two figures, it seems that gold recovered amounts tend to have little increase year after year and the quantities of resources of Au present a small decreased value with time.

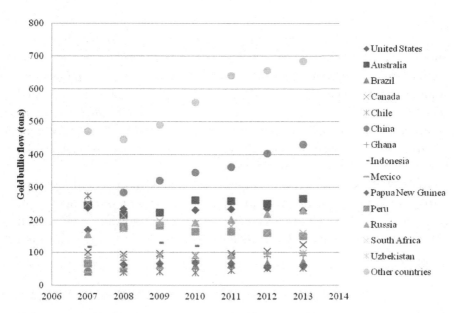

Fig. 1   Gold production worldwide in the last seven years (bullion flow in tons)
Source: USGS, Mineral Commodity Summaries [25]

The Recovery of Gold from Secondary Sources

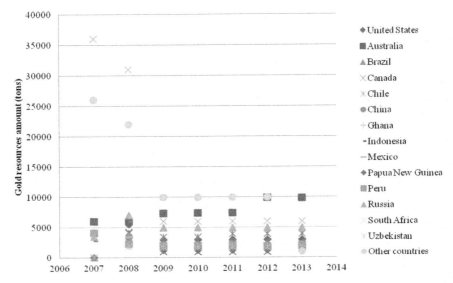

Fig. 2 Evolution of gold resources at global level within the last seven years
Source: USGS, Mineral Commodity Summaries [25]

This noble metal has two oxidation states (+1 — aurous and +3 — auric) and is unaffected by oxygen at any temperature level. As is demonstrated in the Pourbaix diagram depicted in Fig. 3, in simple hydrate solution, gold will never corrode. This is due to the fact that in the presented conditions, the decomposition of water to oxygen without the presence of coordinating ligands will not be enough to oxidize gold even in the presence of either strong acids or strong alkalis. This demonstrates its nobles, Au being the only metal with this property.

This PM will never react in inorganic solution of sulfuric, hydrochloric, nitric, or fluorhidric acids. Also, this does not react with other metals, but reacts with most groups of halides. Moreover, as is shown in Eqs. (1) and (2), gold always needs to be oxidized to accomplish the reaction with a reducing agent.

$$Au \rightarrow Au^+(aq) + e^- \quad (E = 1.7 + \log_{10}[Au^+]), \tag{1}$$

$$Au \rightarrow Au^{3+}(aq) + e^- \quad (E = 1.50 + 0.0197\log_{10}[Au^{3+}]). \tag{2}$$

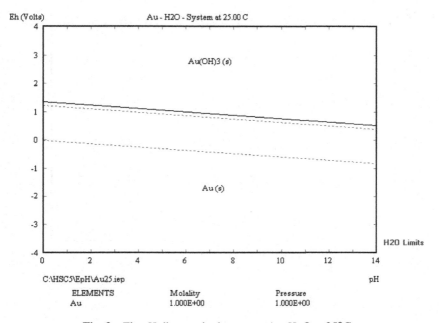

Fig. 3 Eh–pH diagram in the system Au–H₂O at 25°C

7.3 Hydrometallurgical Systems for Gold Recovery from WPCBS

7.3.1 *Cyanidation process for gold recovery*

The use of cyanide was successfully applied in the recovery of Au from minerals since 1887. This compound in the form of NaCN, used generally in a concentration of 0.01% and 0.05% cyanide, is worldwide involved for the treatment of low grade minerals and minerals that cannot be treated by simple physical operations like churching and gravity separation [26]. The main involved reaction during the leaching process, also called the Elsner Equation, is

$$4Au + 8CN^- + O_2 + 2H_2O = 4Au(CN)_2^- + 4OH^-. \qquad (3)$$

As can be seen in Fig. 4, gold forms a stable complex with cyanide within the water stability lines. The dissolution of Au with cyanide takes

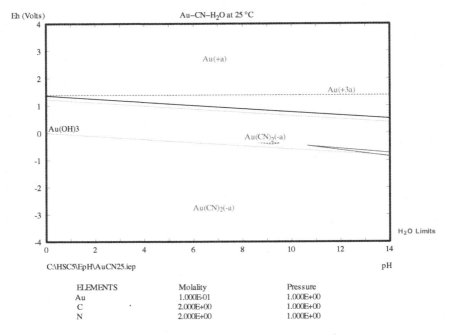

Fig. 4   Eh–pH diagram for Au–CN–H₂O at 25°C

place if the following conditions are ensured [6]: at a pH value of about 9, as can be seen in Fig. 4, the gold forms a stable complex with cyanide; the decrease in this value leads to the formation of toxic hydrogen cyanide. This is explained by the presence of more than 90% of the total added cyanide which is present as free cyanide and is hydrolyzed if the proper pH value is not ensured; temperature level must be low to avoid the degradation of cyanide; particle sizes must have a certain range as the kinetics of reactions are strongly dependent on this parameter; oxygen concentration; presence of other anions/cations. In the work of Quinet *et al.* [27], a complete flowsheet has been proposed for the treatment of WPCBs. The recovery step of gold was performed after many preliminary steps for the removal of the recovery of base and other PM (Pd and Ag). The cyanidation procedure was considered as a final step of grounded PCBs as the presence of base metals in high concentrations leads to degradation of cyanide reagent. This substance presents a high susceptibility to base metals, particularly for Cu. This is due to the formation of Cu $(CN)_2^-$

that considerably reduces the required amount of cyanide for reaction with Au. According to Muir [28], for each percent of Cu, 30 kg/t of cyanide are consumed. Therefore, this reagent must be continuously added in high excess by the required stochiometric amount in the reaction with gold. In addition, passivation phenomenon may also take place due to the formation of other complexes of cyanide with other base metals. However, as this process has been used successfully for the treatment of mineral ores, in the treatment of e-wastes, this process was replaced with more environment-friendly chemicals, such as thiourea, thiousulphate, thiocyanate, and halogens systems (chlorine–chloride, iodine–iodide, bromine–bromide).

7.3.2 *Thiosulphate gold recovery procedure*

Thiousulfate use in the metallurgical extraction of gold was considered a suitable approach for replacing the cyanide process. This process allows the reduction of interference with external cations and therefore reduction of the environmental impact [29]. The main reaction of gold with thosulfate in the presence of oxygen is presented in Eq. (5).

$$2Au + 0.5O_2 + 4S_2O_2^{-3} + H_2O \rightarrow 2Au(S_2O_3)^{-3}{}_2 + 2OH^-. \quad (5)$$

Even if this process has been intensively investigated for more than 25 years, the chemistry of the process is still very complex, being mostly dependent on simultaneous reactions of copper(II) and oxygen with both gold and thiosulphate (Eqs. (6)–(8)).

$$Au + 2S_2O_3^{2-} \rightarrow Au(S_2O_3)^{3-} + e^-, \quad (6)$$

$$Cu(NH_3)_4^{2+} + 3S_2O_3^{2-} + e^- \rightarrow Cu(S_2O_3)_3^{5-} + 4NH_3. \quad (7)$$

Overall reaction may be expressed as follows:

$$Au + Cu(NH_3)_4^{2+} + 2S_2O_3^{2-}$$
$$= Au(S_2O_3)_2^{3-} + Cu(NH_3)_2^+ + 2NH_3. \quad (8)$$

As presented above, the reaction is electrochemical, with first mechanism of reaction when gold reacts with thiosulfate to form gold thiosulfate

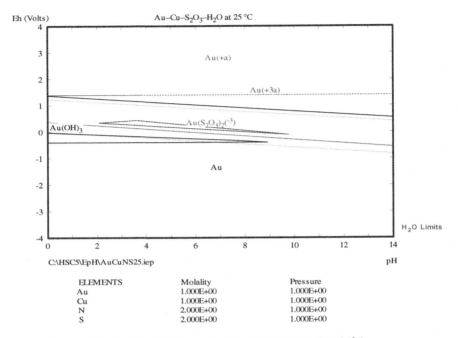

Fig. 5   Eh–pH diagram for Au–Cu–S$_2$O$_3$–H$_2$O at 25°C

(Eq. (6)) and a second one where Cu^{2+} reacts with thiosulfate to form copper thiosulphate (Eq. (7)) [30].

Figure 5 presents the stability Eh–pH diagram for gold leaching with thiosulfate in the presence of copper at 25°C. As presented in the literature, gold forms a very stable complex with thiosulphate within a certain range of pH. Also, for this process, this parameter plays a crucial role in gold recovery. Moreover, the presence of some elements, e.g. iron, can reduce the gold leaching efficiency within this system. The presence of this element in high concentration has a deleterious effect as it degrades the thiosulfate. By reaction with S$_2$O$_3$, iron hydroxide is formed and this further precipitates on the surface of PM, reducing therefore the contact surface of the precious element with the ligand. Another study undertaken by Feng and Van Deventer [31] to study the other base metals influence within this process has shown that only Zn has a negative effect on gold leaching. As presented further, the main inconvenience of this process is the very slow reaction kinetics.

In a study of Ha *et al.* [32], the pretreatment performed on WPCBs of e-scrap and mobile phones has resulted in fast reaction kinetics for the first kind of board material; within 2 h, the authors state that they have achieved a gold leaching efficiency of 98%. Instead, for the mobile phones boards, 10 h of process have been necessary to obtain 90% of Au recovery. The treatment of WPCBs coming from obsolete mobile phones for recovery of gold by thiosulfate leaching has been applied by Tripathi *et al.* [33]. They have conducted their study with variation of different factors like Cu^{+2} concentration, solid concentration, ammonium thiosulfate concentration, and pH value. Under the optimal conditions, they have achieved a gold extraction of about 79%. They have demonstrated that working with a large particle area leads to high copper dissolution degree, thus negatively affecting the reaction mechanism by thiosulfate degradation. These results have also been achieved in another study performed on the mobile phones WPCBs. With the diminishing of particle sizes to less than 0.8 mm, the Au extraction yield has increased from 16% within 48 h to 98% [34].

Considering the large consumption of this reagent, more than 50% of thiousulphate is lost within the leaching procedure [35], even if it is considered a more eco-friendly procedure, this process results in not being economic for the application of in-treatment of WPCBs for Au recovery. In addition, in the recovery step of gold in its solid form, according to the literature data, many other issues are encountered [36, 37] and sometimes it is difficult to achieve high recovery yields for Au.

7.4 Halide Leaching Process of Gold

7.4.1 *Chloride leaching*

The leaching of gold with chloride/chlorine was the only system of halides that has been applied also at the industrial level [38]. The gold reaction with chloride is represented by the following two equations:

$$2Au^{o} + \frac{1}{2}O_2 + 2H^+ + 4Cl^- = 2AuCl_2{}^- + H_2O, \qquad (9)$$

$$2Au^{o} + \frac{3}{2}O_2 + 6H^+ + 8Cl^- = 2AuCl^- + 3H_2O. \qquad (10)$$

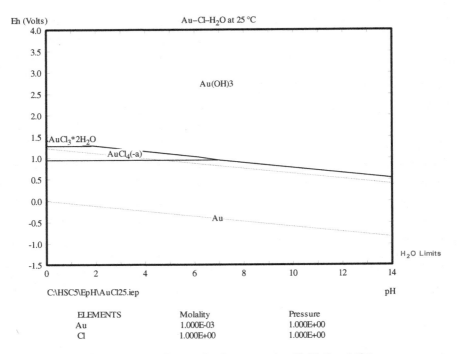

Fig. 6   Eh–pH diagram for the system Au–Cl–H$_2$O at 25°C

The stability of gold complexation with chloride is depicted in Fig. 6. This thermodynamic analysis reveals that oxygen should be enough for Au oxidation; however, as is shown in most of the already tested procedures, only with the uses of strong oxidants like chlorine, permanganate, and hypochlorite, reliable recoveries yields of gold have been achieved. To stabilize the gold complex, most of the studies have reported that the most important aspect of halide systems is the rigorous control of the pH and Eh.

For Au dissolution in chloride solution, chlorine is considered as the strongest catalysis. The kinetics rate of gold dissolution in chlorine media is considered to be much faster than the one with cyanide and as is revealed in Eq. (11), gold chloro-complex ($[Au(III)Cl_4]^-$) is achieved a pH less than 2 [39].

$$2Au^o + 3Cl_2 + 2Cl^- \rightarrow 2[AuCl_4]^-. \qquad (11)$$

The process with hypochlorite as oxidant has attracted interest in the treatment of minerals [39, 40].

$$2Au + 3HClO + 5Cl^- + 3H^+ \rightarrow 2[AuCl_4]^- + 3H_2O. \tag{12}$$

During the leaching process within this system, it was noted that at least 10 g/L OCl^- and 9 g/L HCl are required for a good dissolution efficiency of gold. In addition, it was observed that a decrease of OCL^- leads to precipitation of gold complex into solution, this being a result of the reaction with residual S into the ore [39].

In the patent published by Zhou *et al.* [41], after a thermal treatment of the electronic scrap and two steps carried out to extract base metals and silver, the leaching with hydrochloric acid and sodium hypochlorite was performed for Au recovery. In this way, about 92% of PM efficiency has been achieved with high purity of the final products. For the treatment of WPCBs of mobile phones, Kim *et al.* [42] have used electro-generated chloride from hydrochloric acid media chlorine as leaching system. The authors have noticed that the most important factors that influence this system for the recovery of Au are the Cl^- concentration, temperature, pulp density, and also the presence of Cu. With a large Cu concentration, the level of Au extraction was low as most of the chlorine was consumed to achieve the dissolution of the base metal. For gold leaching, it is also necessary to have a larger solution potential. Therefore, to achieve good process selectivity, the process was divided into two steps where firstly copper was recovered. In this way, 93% of Au extraction has been achieved.

7.4.2 Iodide/iodine leaching of Au

Gold dissolution with iodide/iodine system is also considered less toxic than the treatment with cyanide. The complex of gold is considered to be less stable than the cyanide one but stronger than the complexes realized with chlorine, bromine, thiocyanates, and cyanides. The main process reactions in a I^-/I_3^- system are [43] as follows:

$$2Au + I^- + I_3^- = 2AuI_2^-, \tag{13}$$

$$2Au + 3I_3^- = 2AuI_4^- + I^-. \tag{14}$$

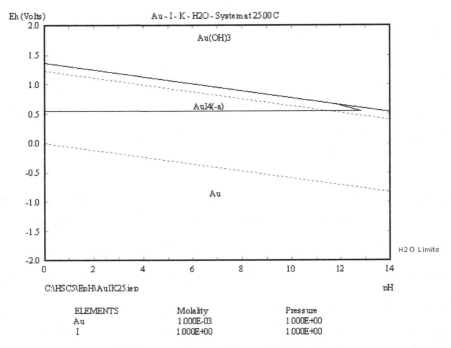

Fig. 7   Eh–pH diagram for Au–I–H$_2$O at 25°C

The thermodynamics of gold leaching with iodide/iodine system from Fig. 7 shows that the AuI$_4^-$ complex has a large interval of stability. This aspect was studied and confirmed by many researchers. They have concluded that the kinetics of gold leaching remain unaffected over large pH value [44–46]. This aspect results in a lower consumption of iodine when this is compared with the cyanide. Furthermore, with these advantages, iodide/iodine system may find application in so-called solution mining of gold process [47].

Until now, the gold leaching into this system is believed to be a high cost process due to the high price of iodine [41]. Thus, this approach has not yet found high interest for its application in the treatment of WPCBs. This is demonstrated by the scarcity of the scientific literature data in the treatment of these wastes with iodine leaching procedure.

Xiu *et al.* [48] have proposed the use of this leaching procedure on the WPCB of spent mobile phones. For this reason, to achieve the optimal conditions, an intensive study was performed after the supercritical treatment of the waste material and leaching step for recovery of base metals. As in the case of the minerals, as is mentioned before, it is required to avoid the consumption of iodine, precipitation of gold complex due to the reaction of iodine with the other elements that are present in the composition of the waste board and also to obtain high grade products. At the end of the procedure, more than 95% of gold extraction yields have been achieved.

A main conclusion of this process may say that it can be considered as environment friendly, with high levels of recoveries, there is no economy in the application of this system in the treatment of e-wastes. Like in the case of chlorine system, the investment cost of required devices for the treatment and also the high price of the chemicals represent the main drawbacks of this technology.

7.4.3 *Bromide/bromine leaching process of Au*

Bromine has been introduced into the chemistry as solvent for gold solubilization process in 1846 [49]. Even if this chemical was introduced many years ago, there are not much scientific data in the literature about this process. The process of gold dissolution with bromine in the presence of bromide and chloride gas has been patented by Fink and Putnam. As revealed in Fig. 8, gold metal may be oxidized to form in the acidic area both $AuBr_2$ and $AuBr_4$ complexes. The complexes of gold with bromide have an intermediary stability between the complexes that Au forms with iodide and chloride.

The leaching of Au with this chemical presents as main characteristics the good fast rate of reaction kinetics and application from the acidic to the neutral level of pH (Fig. 8).

In the case of bromine used as oxidizing agent for gold, the system reaction is as follows:

$$2Au + 3Br_2 + 2Br^- = 2AuBr_4^-. \tag{15}$$

It has been observed that in the mineral ores, this treatment may be advantageous; no further step of neutralization is required for further metal extraction from solution. However, if some compounds like sulfide are present, this negatively affects the process economy as a large consumption of Br is realized [50].

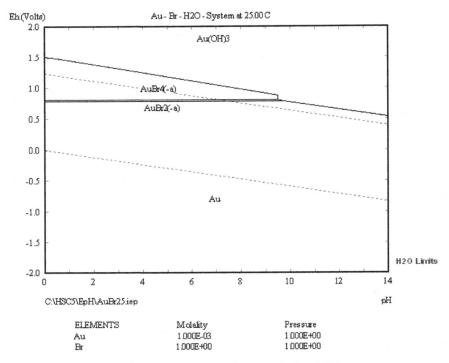

Fig. 8 Eh–pH diagram of Au–Br–H$_2$O at 25°C

The treatment of WPCBs with bromide/bromide system after two leaching step for the recovery of base metals has been patented by Kogan [51]. The first two leaching consisted of using hydrochloric acid and, respectively, sulfuric acid as solvents. In both leaching systems, magnesium chloride was used as oxidant. The leaching of Au undertaken by the solid residues of both previous solubilization procedures has been performed with sodium chloride in an acid media of hydrochloric acid. This procedure has resulted in Au recovery degree larger than 98%.

Until now, there are no other scientific approaches on the treatment of WPCBs with this system.

7.5 Thioureation Process for Au Recovery

Thiourea or thiocarbimine is a chemical substance with low toxicity and it was extensively studied by many researchers due to its very fast reaction

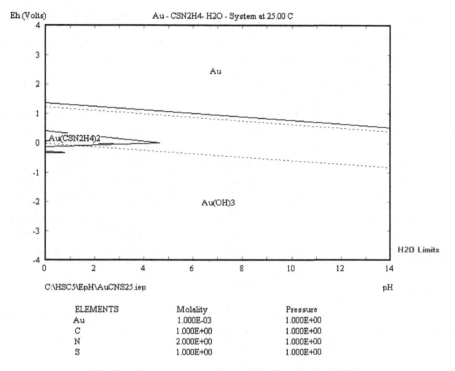

Fig. 9   Eh–pH diagram of Au–CSN$_2$H$_4$–H$_2$O at 25°C

rate of complexation with gold forming a stable cationic complex (Au I), and gold extraction larger than 99% can be achieved. The reaction of gold with CSN$_2$H$_4$ takes place with a pH interval of 1–2 (Fig. 9) and is expressed by the following equation:

$$Au + 2CSN_2H_4 = Au(CSN_2H_4)_2 + e^-. \tag{16}$$

The most effective system of the gold leaching in thiourea acid solution is considered to be the one with triferric ion as oxidant. According to Li and Miller [52], the process is strongly dependent on the solution potential and the agitation rate. In addition, the particle size also plays a crucial role in the dissolution of Au into this system [53–55]. The reaction of gold with thiourea in sulfuric acid media in the presence of ferric sulfate as oxidant is described by the reaction in Eq. (17). However, the oxidation of thiourea by ferric ion leads to an irreversible reaction with formamidine disulfide (FDS) formation, a complex which is further decomposed to elemental sulfur and cyanamide (Eq. (18)).

$$2Au + 2Fe^{3+} + 4CSN_2H_4 + SO4^-$$
$$\rightarrow 2[Au(CSN_2H_4)_2]SO_4 + 2Fe^{2+} \qquad (17)$$

$$H_2N-CNH-S-S-CNH-NH_2 \rightarrow H_2N-CS-NH_2 + H_2N-C \equiv N + S.$$
$$(18)$$

The thioureation procedure is considered more advantageous to the cyanide process due to its less sensibility to base metals (Cu, Pb, Zn) and lower toxicity. However, it was noted that a large concentration of Cu in the material with Au content determines the degradation of thiourea to glutinous elemental sulfur which very fast forms passivation layer on the PM surface. As reported in Eh–pH diagram (Fig. 10) for the thermodynamics of Au and

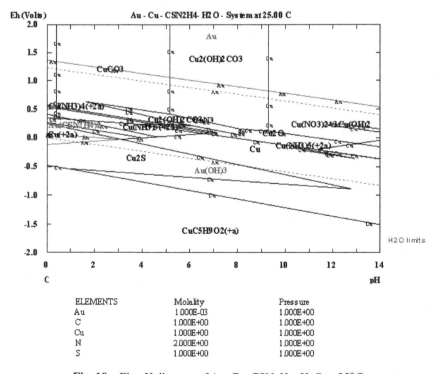

ELEMENTS	Molality	Pressure
Au	1.000E-03	1.000E+00
C	1.000E+00	1.000E+00
Cu	1.000E+00	1.000E+00
N	2.000E+00	1.000E+00
S	1.000E+00	1.000E+00

Fig. 10   Eh–pH diagram of Au–Cu–CSN$_2$H$_4$–H$_2$O at 25°C

Cu reaction with thiourea at a relative high concentration of copper (1M/L), thiourea reacts with copper to form the cupric complex that further catalyzes the thiourea oxidation by ferric ion [55–57].

In the treatment of WPCBs by acid thiourea leaching procedure, as the copper concentration has a range of 16–20% or even larger, there is always a preliminary step required for the complete recovery of this base metal. For this reason, many authors have applied preliminary oxidative leaching process [55, 57–61]. Another important factor that can influence the gold solubilization was the particle size. In our study [55], the diminishing of particle dimension from 3 mm to less than 2 mm the gold recovery yield was increased from 45% to about 70%. This result was also achieved in the study of gold recovery from WPCB of exhausted mobile phones of Jing-Ying *et al.* [54]. In addition, by continuing the experimental work, after more than 98% of Cu removal from WPCBs, sample which had a dimension of less than 1 mm, approximately all gold recovery has been achieved. As a large quantity of thiourea was used in the treatment of these wastes, a three-step cross-leaching procedure has been performed. This was performed by using the solution three times with the indication that a smaller amount than in the first step has been added in the second and third steps. This is explained by the consumption of the thiourea by its reaction with ferric ion. Therefore, a supplementary addition of CSN_2H_4 will provide the required amount for both gold transfer into solution and enrichment of this PM content into solution [57].

7.6 Conclusions

In this chapter, current technologies for the recovery of Au from WPCBs by solubilization procedure have been presented. The main factor influences during all the processes have been also addressed. Each procedure has revealed that a good control of the process parameters and a pretreatment of the gold containing materials positively affect the dissolution rate and efficiency. It was noted that in the scientific literature, a high interest in the replacing of cyanide process was focused in the thioureation procedure. However, it is still necessary to continue the process optimization for the reduction of large consumption of this chemical.

References

[1] Hadi P, Xu M, Lin CSK, Hui CW, McKay G (2015) Waste PCB recycling techniques and product utilization. *Journal of Hazardous Materials*, **283**: 234–243. 10.1016/j.jhazmat.2014.09.032.

[2] Zoeteman BCJ, Krikke HR, Venselaar J (2010) Handling WEEE waste Hows: On the effectiveness of producer responsibility in a globalizing world. *International Journal of Advanced Manufacturing Technology*, **47**(5–8): 415–436.

[3] Wang X, Gaustad G (2012) Prioritizing material recovery for end-of-life PCBs. *Waste Management*, **32**(10): 1903–1913.

[4] Bizzo WA, Figueiredo RA, De Andrade VF (2014) Characterization of PCBs for metal and energy recovery after milling and mechanical separation. *Materials*, **7**(6): 4555–4566.

[5] Prismark (2012) Retrieved November 1, 2015, from http://www.prismark.com/what_publications_pcr.php.

[6] Akcil A, Erust C, Gahan CS, Ozgun M, Sahin M, Tuncuk A (2015) Precious metal recovery from waste PCBs using cyanide and non-cyanide lixiviants — A review. *Waste management (New York, N.Y.)*, **45**: 258–271. 10.1016/j.wasman.2015.01.017.

[7] GAIA (Global Alliance for Incinerator Alternatives) (2013) Waste Incinerators?: Bad News for Recycling and Waste Reduction, (October), 1–9. Retrieved from www.no-burn.org.

[8] Hagelüken C (2005) Recycling of electronic scrap at Umicore's integrated metals smelter and refinery. *Proceedings of EMC*, **59**(3): 1–16.

[9] Hagelüken C, Corti CW (2010) Recycling of gold from electronics: Cost-effective use through "Design for Recycling." *Gold Bulletin*, **43**(3): 209–220.

[10] Long L, Sun S, Zhong S, Dai W, Liu J, Song W (2010) Using vacuum pyrolysis and mechanical processing for recycling waste PCBs. *Journal of Hazardous Materials*, **177**(1–3): 626–632. 10.1016/j.jhazmat.2009.12.078.

[11] Hall WJ, Williams PT (2007) Separation and recovery of materials from scrap PCBs. *Resources, Conservation and Recycling*, **51**: 691–709. 10.1016/j.resconrec.2006.11.010.

[12] Zhou Y, Wu W, Qiu K (2010) Recovery of materials from waste PCBs by vacuum pyrolysis and vacuum centrifugal separation. *Waste management (New York, N.Y.)*, **30**(11): 2299–2304. 10.1016/j.wasman.2010.06.012.

[13] Guo Q, Yue X, Wang M, Liu Y (2010) Pyrolysis of scrap PCB plastic particles in a fluidized bed. *Powder Technology*, **198**(3): 422–428. 10.1016/j.powtec.2009.12.011.

[14] Chien YC, Paul Wang H, Lin KS, Huang YJ, Yang Y (2000) Fate of bromine in pyrolysis of PCB wastes. *Chemosphere*, **40**(4): 383–387. 10.1016/S0045-6535(99)00251-9.

[15] Mankhand TR, Singh KK, Kumar Gupta S, Das S (2013) Pyrolysis of PCBs. *International Journal of Metallurgical Engineering*, **1**(6): 102–107. 10.5923/j.ijmee.20120106.01.

[16] Flandinet L, Tedjar F, Ghetta V, Fouletier J (2012) Metals recovering from waste PCBs (WPCBs) using molten salts. *Journal of Hazardous Materials*, **213–214**: 485–490. 10.1016/j.jhazmat.2012.02.037.

[17] Li J, Liang C, Ma C (2015) Bioleaching of gold from waste PCBs by Chromobacterium violaceum. *Journal of Material Cycles and Waste Management*, **17**(3): 529–539. 10.1007/s10163-014-0276-4.

[18] Chi TD, Lee JC, Pandey BD, Yoo K, Jeong, J (2011) Bioleaching of gold and copper from waste mobile phone PCBs by using a cyanogenic bacterium. *Minerals Engineering*, **24**(11): 1219–1222. 10.1016/j.mineng.2011.05. 009.

[19] Faramarzi MA, Stagars M, Pensini E, Krebs W, Brand H (2004) Metal solubilization from metal-containing solid materials by cyanogenic Chromobacterium violaceum. *Journal of Biotechnology*, **113**: 321–326. 10.1016/j.jbiotec.2004.03.031.

[20] Tay, SB, Natarajan G, Rahim MNBA, Tan HT, Chung MCM, Ting YP, Yew WS (2013) Enhancing gold recovery from electronic waste via lixiviant metabolic engineering in Chromobacterium violaceum. *Scientific reports*, **3**: 2236. 10.1038/srep02236.

[21] Kumar Pradhan J, Kumar S (2012) Metals bioleaching from electronic waste by Chromobacterium violaceum and Pseudomonads sp. *Waste Management & Research*, **30**(11): 1151–1159. 10.1177/0734242X12437565.

[22] Dangton J, Leepowpanth Q (2014) A study of gold recovery from e-waste by bioleaching using *Chromobacterium violaceum*. *Applied Mechanics and Materials*, **548–549**: 280–283. 10.4028/www.scientific.net/AMM.548-549.280.

[23] Ruan J, Zhu X, Qian Y, Hu J (2014) A new strain for recovering PM from waste PCBs. *Waste Management*, **34**(5): 901–907. http://dx.doi.org/ 10.1016/j.wasman.2014.02.014.

[24] Jujun R, Jie Z, Jian H, Zhang J (2015) A novel designed bioreactor for recovering pm from waste PCBs. *Scientific Reports*, **5**: 13481. 10.1038/srep13481.

[25] USGS Minerals information (2015) Mineral Commodity Summaries. Retrieved from http://minerals.usgs.gov/minerals/pubs/mcs/.

[26] Logsdon MJ, Hagelstein K, Mudder TI (2001) *The management of cyanide in gold extraction.* Retrieved from http://www.icmm.com/page/1616/ the-management-of-cyanide-in-gold-extraction.

[27] Quinet P, Proost J, Van Lierde a (2005). Recovery of PM from electronic scrap by hydrometallurgical processing routes. *Minerals and Metallurgical Processing*, **22**(1): 17–22. Retrieved from http://www.scopus.com/inward/record.url?eid=2-s2.0-16244393080&partnerID=40&md5=d3a365564741 7a2562bd86a5f099166a.

[28] Muir DM (2011) A review of the selective leaching of gold from oxidised copper-gold ores with ammonia-cyanide and new insights for plant control and operation. *Minerals Engineering*, **24**(6): 576–582. 10.1016/j.mineng.2010.08.022.

[29] Abbruzzese C, Fornari P, Massidda R, Vegliò F, Ubaldini S (1995) Thiosulphate leaching for gold hydrometallurgy. *Hydrometallurgy*, **39**(1–3): 265–276. 10.1016/0304-386X(95)00035-F.

[30] Breuer PL, Jeffrey MI (2000) Thiosulphate leaching kinetics of gold in the presence of copper and ammonia. *Minerals Engineering*, **13**(10): 1071–1081.

[31] Feng D, Van Deventer JSJ (2002) The role of heavy metal ions in gold dissolution in the ammoniacal thiosulphate system. *Hydrometallurgy*, **64**(3): 231–246. 10.1016/S0304-386X(02)00046-4.

[32] Ha VH, Lee JC, Jeong J, Hai HT, Jha MK (2010) Thiosulfate leaching of gold from waste mobile phones. *Journal of Hazardous Materials*, **178**(1–3): 1115–1119. 10.1016/j.jhazmat.2010.01.099.

[33] Tripathi A, Kumar M, Sau DC, Agrawal A, Chakravarty S, Mankhand TR (2012) Leaching of gold from the waste mobile phone PCBs with ammonium thiosulphate. *International Journal of Metallurgical Engineering*. Scientific & Academic Publishing. 10.5923/j.ijmee.20120102.02.

[34] Ficeriová J, Baláž P, Gock E (2011) Leaching of gold, silver and accompanying metals from circuit boards (PCBs) waste. *Acta Montanistica Slovaca*, **16**(2): 128–131. Retrieved from http://www.scopus.com/inward/record.url?eid=2-s2.0-84856561238&partnerID=tZOtx3y1.

[35] Zhang Y, Liu S, Xie H, Zeng X, Li J (2012) Current status on leaching precious metals from waste PCBs. *Procedia Environmental Sciences*, **16**: 560–568. 10.1016/j.proenv.2012.10.077.

[36] Fotoohi B, Mercier L (2014) Recovery of precious metals from ammoniacal thiosulfate solutions by hybrid mesoporous silica: 1-Factors affecting gold adsorption. *Separation and Purification Technology*, **127**: 84–96. 10.1016/j.seppur.2014.02.024.

[37] Hiskey JB, Lee J (2003) Kinetics of gold cementation on copper in ammoniacal thiosulfate solutions. *Hydrometallurgy*, **69**(1–3): 45–56. 10.1016/S0304-386X(03)00003-3.

[38] Bünyamin Dönmez, Fatih Sevim SÞ (2001) Study on recovery of gold from decopperized anode slime. *Chemical Engineering & Technology*, **24**: 91–95.

[39] Baghalha M (2007) Leaching of an oxide gold ore with chloride/hypochlorite solutions. *International Journal of Mineral Processing*, **82**(4): 178–186. 10.1016/j.minpro.2006.09.001.

[40] Nam KS, Jung BH, An JW, Ha TJ, Tran T, Kim MJ (2008) Use of chloride-hypochlorite leachants to recover gold from tailing. *International Journal of Mineral Processing*, **86**: 131–140. 10.1016/j.minpro.2007.12.003.

[41] Syed S (2012) Recovery of gold from secondary sources — A review. *Hydrometallurgy*, **115–116**: 30–51. 10.1016/j.hydromet.2011.12.012.

[42] Kim EY, Kim MS, Lee JC, Pandey BD (2011) Selective recovery of gold from waste mobile phone PCBs by hydrometallurgical process. *Journal of Hazardous Materials*, **198**: 206–215. 10.1016/j.jhazmat.2011.10.034.

[43] Baghalha M (2012) The leaching kinetics of an oxide gold ore with iodide/iodine solutions. *Hydrometallurgy*, **113–114**: 42–50. 10.1016/j.hydromet.2011.11.013.

[44] Davis A, Tran T (1991) Gold dissolution in iodide electrolytes. *Hydrometallurgy*, **26**(2): 163–177. 10.1016/0304-386x(91)90029-L.

[45] Qi PH, Hiskey JB (1991). Dissolution kinetics of gold in iodide solutions. *Hydrometallurgy*, **27**: 47–62.

[46] Qi PH, Hiskey JB (1993) Electrochemical behavior of gold in iodide solutions. *Hydrometallurgy*, **32**(2): 161–179. 10.1016/0304-386X(93)90021-5.

[47] Kubo S (1994) Development of gold ore leaching method by iodine. In *In Situ Recovery of Minerals II Santa Barbara California USA 2530 Oct 1992* (pp. 405–431). Engineering Foundation.

[48] Xiu FR, Qi Y, Zhang FS (2015) Leaching of Au, Ag, and Pd from waste PCBs of mobile phone by iodide lixiviant after supercritical water pre-treatment. *Waste Management*, **41**: 134–141. 10.1016/j.wasman.2015.02.020.

[49] Kuzugüdenli OE, Kantar Ç (1999) Alternates to gold recovery by cyanide leaching. *Erciyes Üniversitesi Fen-Ed. Fakültesi*, **2**: 119–127. Retrieved from http://fbe.erciyes.edu.tr/Turkce/eufbedergisi/DER99/119-127.pdf.

[50] Melashvili M, Fleming C, Dymov I, Manimaran M, O'Day J (2014) Study of Gold Leaching With Bromine and Bromide and the Influence. *Conference of Metallurgists Proceedings ISBN: 978-1-926872-24-7*, (November 2015).

[51] Kogan V (2006) Process for the Recovery of Precious Metals from Electronic Scrap by Hydrometallurgical Technique, Int. Patent WO/2006/013568.

[52] Li J, Miller JD (2007) Reaction kinetics of gold dissolution in acid thiourea solution using ferric sulfate as oxidant. *Hydrometallurgy*, **89**(3–4): 279–288. 10.1016/j.hydromet.2007.07.015.

[53] Ficeriová J, Baláž P (2010) Leaching of gold from a mechanically and mechanochemically activated waste. *Acta Montanistica Slovaca*, **15**(3): 183–187.

[54] Jing-ying L, Xiu-li X, Wen-quan L (2012) Thiourea leaching gold and silver from the PCBs of waste mobile phones. *Waste Management*, **32**(6): 1209–1212.

[55] Birloaga I, De Michelis I, Ferella F, Buzatu M, Vegliò F (2013) Study on the influence of various factors in the hydrometallurgical processing of waste PCBs for copper and gold recovery. *Waste Management*, **33**(4): 935–941.

[56] Krzewska S, Podsiadly H, Pajdowski L (1980) Studies on the reaction of copper(II) with thiourea. III. Equilibrium and stability constants in copper(II)-thiourea-perchloric acid redox system. *Journal of Inorganic and Nuclear Chemistry*, **42**: 89–94. Retrieved from internal-pdf://krzewska_et_al_1980-0664689920/Krzewska_et_al_1980.pdf.

[57] Birloaga I, Coman V, Kopacek B, Vegliò F (2014) An advanced study on the hydrometallurgical processing of waste computer PCBs to extract their valuable content of metals. *Waste Management*, **34**(12): 2581–2586. Retrieved from http://www.sciencedirect.com/science/article/pii/S0956053X14003833.

[58] Ionela Birloaga, Francesco Vegliò, Ida DeMichelis BK (2014) Hydrometallurgical processing of waste PCBs for Cu, Au and Ag recovery. In *Towards a Resource Efficient Economy*. Vienna.

[59] Kamberovic Z, Korac M, Ivsic D, Nikolic V, Ranitovic M (2009) Hydrometallurgical process for extraction of metals from electronic waste, Part II: development of the processes for the recovery of copper from PCBs (PCB). *Metallurgical & Materials Engineering*, **15**(4): 231–243.

[60] Behnamfard A, Salarirad MM, Veglio F (2013) Process development for recovery of copper and precious metals from waste PCBs with emphasize on palladium and gold leaching and precipitation. *Waste Management*, **33**(11): 2354–2363. 10.1016/j.wasman.2013.07.017.

[61] Camelino S, Rao J, Padilla RL, Lucci R (2015) Initial studies about gold leaching from printed circuit boards (PCB's) of waste cell phones. *Procedia Materials Science*, **9**: 105–112. 10.1016/j.mspro.2015.04.013.

Index

A

activated carbon, 108
activation, 111
activation temperature, 127
active iodine, 48
adsorption, 145
adsorption capacity, 113
adsorption kinetics, 117
alluvial deposits, 8
amalgamation, 99
amine complexes, 16
anionic degradation, 22
aqua regia, 81
aromatic, 112
Au resources, 201
Au–Ni alloy, 10, 40
auric chloride, 37
aurocyanide, 107
aurocyanide complex, 100
aurothiosulfate complex, 20
aurous complex, 38
autocatalytic electroless deposition, 58

B

ball-pan method, 121
band diagram of silicon, 72
batch experiments, 120
bio-oxidation, 101
bioadsorbents, 146
bioleaching, 182

biopolymers, 145
burnout, 112

C

carbon, 107
carbon granule, 127
carbon-in-column, 104
carbon-in-leach, 104
carbon-in-pulp, 104
carbonaceous precursor materials, 122
carbonization, 111
catalytic nuclei, 76
cellulose, 151
cementation of gold, 28, 68
cementation of gold using copper, 68
cementation of gold using zinc, 67
cementation processes of gold, 58
cementation using aluminum, 68
challenges, 27, 34, 43, 49
chemical control, 25
chemical oxidation, 101
chemically controlled, 33
chemistry, 205
chloride, 177–178, 180–181, 183–186
chlorination, 100
chlorine adsorption, 44
chlorine generation, 36
chlorine species, 40
Cl^- ions concentration, 42
coconut shells, 116

competitive reactions, 80
complex, 107
conditioning, 122
constant (k), 119
copper concentration, 24
corrosive, 44
cracking pressure, 125
Cu(II)–Cu(I) ammine complexes, 19
Cu(II)–thiosulfate complex, 19
current density, 43
cyanidation, 35
cyanidation process, 102
cyanide, 174, 179, 181–182, 187–191
cyanide process, 100

D

defect detection of silicon, 58
degradation products, 31
dehydration, 110
diaminoaurate(I) complex, 19
diffusion control, 24
diffusion rate, 41
displacement deposition, 77
displacement metal deposition, 61
dissolution, 203
dissolved chlorine, 38
Dubinin, 113

E

e-waste, 8, 16
EDX analysis, 20
Eh–pH diagram, 17, 206
electrical conductivity, 97
electrochemical, 32
electrochemical mechanism, 39
electrochemical reactions, 18, 45
electroless displacement deposition,
    57–59, 62, 65, 68–69, 76, 79, 81
electrolytic cell, 36
electrolytic mechanism, 20
electromotive force (ΔE), 61, 63
electronegativity, 44
electronic devices, 197
electronic wastes, 198
electrowinning, 106

equilibrium constant (K), 119
extraction, 32

F

ferric ions, 31
film diffusion, 118
flotation, 99
formamidine disulfide, 29
forms of activated carbon, 109
fractures, 98

G

gold, 7, 96, 199
gold cementation reaction using zinc from
    ammoniacal thiosulphate, 68
gold chloro-complex, 37
gold leaching, 40, 210
gold ores, 98
gold particles, 148
gold passivation, 23
gold recovery from WPCBs, 203
gold–iodide complexes, 44
gold–cyanide complex, 13
gold–thiourea leaching, 30
graphitic layers, 107
gravity concentration, 99
greener alternatives, 13
greener lixiviant, 14

H

H-carbons, 115
half-cell reaction, 33
halides, 35, 180–181, 183
halogens, 35
heap leaching, 105
high resistance, 116
high stability of hydrogen termination,
    79
highly orientated pyrolytic graphite, 119
hole injection, 79
hydrolysis decomposition, 29
hydrometallurgical procedure, 200
hydrometallurgical processing, 102
hydrometallurgical technologies, 144
hydrothermal deposits, 98

I

industrial wastes, 76
influence of aqua regia, 80
iodine concentration, 48
iodo-complex, 45
ion-exchange resins, 107

K

kinetics, 25

L

L-carbons, 115
leaching, 174, 176, 178–191
    bacterial leaching, 12
    chemical leaching, 12
    chlorination leaching, 36
    chlorine–chloride leaching, 14
    cyanide leaching, 11
    gold production, 13
    gold–thiosulfate–ammonia leaching,
      18
    gold–thiosulfate leaching, 22
    iodide leaching, 47
    iodine leaching, 44
    oxidative leaching, 35
    thiourea leaching, 28
leaching process(es), 99, 203
leaching rate, 40
lignin, 146
lignophenol, 146
lime, 102
limitations, 27, 34, 43, 49
lixiviant, 10
    acidic, 11
    ammonia, 18
    ammoniacal, 11
    aqua regia, 12
    cyanide, 11, 13
    sulfuric acid, 35
    thiosulfate, 18
local anode reaction, 59
local cathode reaction, 59
local galvanic cell, 59
low catalytic activity for hydrogen
    evolution, 79

low cost, 85, 87
low environmental impact, 85, 87

M

macadamia nuts, 124
macropores, 114
mechanisms, 32, 118
melting temperature, 96
mercury, 173–174
Merrill–Crowe process, 57, 66–67, 106
mesopores, 114
metal assisted etching of silicon, 58
metal impurity affecting silicon
    semiconductor device, 58
micropores, 114
microporous structure, 108
mobile phone-printed circuit boards
    (MP-PCBs), 16, 20

N

Nernst boundary layer, 38
Nernst equation, 45, 60
Ni(II)–ammine complex, 20
noble metals, 58
noble metal catalysts on silicon, 58
noble metal displacement deposition, 76
noble metal nanoparticles, 58
noble metal selectivity, 76–77
non-corrosive, 49
non-toxic, 49
nucleation activity, 76
nut shells and stones, 123
nut-shell sources, 109

O

ORP, 43
oxidants, 29
oxidation, 59
oxidation mechanism, 103
oxidation of metals, 31
oxidation states of gold, 102
oxidizing agent, 32

P

panning, 99
paper and cotton, 155

parameters, 123
particle size, 85
passivation, 18, 49
passivation layer, 104
printed circuit boards (PCBs), 10, 144
peat, 116
persimmon, 155
photoelectrochemical solar cells, 58
polyphenol, 155
polysaccharides, 151
polythionates, 21
pore diffusion, 118
pore structuring, 127–128
pore widening, 112
potential of silicon energy band, 79
precious metals, 10, 198
pretreatment, 47
printed wiring boards, 198
processing, 123
    aqueous processing, 11
    carbon-in-pulp (CIP), 13
    cyanidation, 13
    hydrometallurgical processing,
        10
    pyrometallurgical route, 10
pure gold, 96
pure gold powder, 74, 83
pyrometallurgical, 174

R

rate constant, 119
rate determining step, 38
rate of gold holding, 121
rate of gold leaching, 26
rate-limiting reaction, 33
reaction Gibbs energy, $\Delta_r G$, 61, 63
reagents, 14
recycling, 8, 21, 198
recycling cycle, 85
redox, 31
redox couple, 17
reduction, 59
reduction of gold(III) ion, 150
refractory ores, 101
removal of volatiles, 110

renewable precursors, 123
retarding effect, 42

S

sawdust silicon, 71, 85
scanning tunnelling microscopy, 120
secondary source, 84
self-catalytic behavior, 16
semiconductor industries, 85
sequence of noble metal recycling, 86
sheets, 112
silicon valence band, 79
single-crystalline silicon, 70
    amorphous, 70
    for multi-crystalline wafers, 70
    micro-crystalline, 70
solar hydrogen evolution, 58
sorption loss, 28
spent mobile phones, 159
spent MP-PCBs, 14
stability Eh–pH diagram, 206
standard electrode potential(s), 60, 63, 65,
    69, 72
stream placer, 98
sulfur-based thio-compounds, 14
supercritical water oxidation (SCWO), 47
surface area of silicon, 76
surface chemistry, 114
surface diffusion, 118
surface reaction, 76
surface reactivity, 108
surface remodeling, 127
surface-enhanced infrared absorption
    reflection spectroscopy, 58
sustainability, 8

T

tannin, 155
tenorite, 25
thermodynamic analysis, 208
thiocyanate, 180–182, 191
thiosulfate, 179–180, 182, 187, 189
thiosulfate concentration, 21
thiosulfate leaching, 16
thiourea, 177, 180–182, 188–191, 213
thiourea concentration, 30

thioureation process, 212
total surface area, 128
tri-iodide ions, 45
typical ores of urban mines, 85

U

urban mines, 58, 76
urea, 30

W

waste electrical and electronic equipment
(WEEE), 58, 144, 173, 175, 181–182

waste hydrofluoric acid solution, 85
waste IC chips, 82–83
waste of wafering or dicing of silicon, 84
waste printed wiring boards (WPCBs),
200
waste silicon powder, 69, 84
waste solution of electroplating, 82
wet attrition, 121

Z

zero-order kinetics, 27

Printed in the United States
By Bookmasters